CLIMATE GAMBLE

Rauli Partanen
Janne M. Korhonen

Climate Gamble

Is Anti-Nuclear Activism
Endangering Our Future?

*"When the facts change,
I change my mind.
What do you do, sir?"*

John Maynard Keynes

CLIMATE GAMBLE –
Is Anti-Nuclear Activism Endangering Our Future?
Rauli Partanen & Janne M. Korhonen © 2015

This book is based on the original
*Uhkapeli ilmastolla -
vaarantaako ydinvoiman vastustus maailman tulevaisuuden?*,
published in Finnish in 2015.

Second, revised edition

Also available through CreateSpace, An Amazon.com Company.
For larger orders, please contact the authors directly.

ISBN 978-952-7139-06-6 (book)
ISBN 978-952-7139-07-3 (ebook)

COVER ILLUSTRATION:
Katariina Pekkola

LAYOUT:
Viestintätoimisto CRE8 Oy
Helsinki 2015

TABLE OF CONTENTS

Foreword

Climate change action is remarkably difficult. Society has many levers available:

√ *demand-reduction through lifestyle change or technology changes;*
√ *eating less meat;*
√ *bioenergy;*
√ *wind power;*
√ *solar power;*
√ *hydro-electricity;*
√ *carbon capture and storage;*
√ *nuclear power;*
√ *carbon-dioxide removal;*
√ *reforestation;*
√ *solar radiation management;*
√ *population reduction.*

Every lever has technical limits and political difficulties. Bioenergy, for example, requires very large land areas, and may have environmental impacts. Eating less meat could make an enormous impact [see globalcalculator.org], but many view vegetarianism as a political non-starter. Some people object to the land area required for wind power and solar power, and the intermittency of wind and solar is a technical challenge.

Making a plan that adds up and that is politically and economically credible is not easy.

Anyone who suggests that one of these levers should not be used by society must recognise that this constraint inevitably makes the task of climate change action harder.

I think that some people view nuclear power as untouchable because the language for describing the dangers of nuclear radiation is too black and white. When we talk about other forms of radiation, everyone understands that there is a scale ranging from harmful to harmless, and we have nuanced language to distinguish between, for example "desert sunshine" and "moonlight", and other levels of "bright" or "dim" radiation in-between.

Everyone knows that midday desert sun can be harmful if one lies in it without protection. And everyone knows that moonlight is essentially harmless. Yet moonlight is made up of just the same photons as sunshine! The reason why moonlight is harmless is that it is 400,000 times less bright than sunshine.

When people talk about nuclear radiation, our language lacks analogous terms for "bright" and "dim". Nuclear radiation is just said to be "toxic", "harmful" or "dangerous". Black and white. But in fact nuclear radiation can be like sunlight, and it can be like moonlight. There are levels of radiation that are lethal, and levels of radiation that are essentially harmless.

Responsible citizens should not simply rule out nuclear power from the portfolio of climate-change options without properly, quantitatively understanding the true risks.

Yes, the nuclear industry has had accidents. Yes, in some countries, the nuclear industry has had a reprehensible track-record of mis-management and dishonesty. Yes, nuclear waste lasts a long time (as do many environmental pollutants).

But please don't leap to simplistic conclusions.

We owe it to our children to behave like adults.

Have an open mind.

Read. Learn. Think.

Discuss.

David J C MacKay FRS[*]
djcm1@cam.ac.uk
Regius Professor of Engineering
Cambridge University Engineering Department

---

[*] Author of "Sustainable Energy - without the hot air" <www.withouthotair. com> and "Information Theory, Inference, and Learning Algorithms" <www.inference.eng.cam.ac.uk/mackay/>. Former Chief Scientific Advisor, Department of Energy and Climate Change, UK.

Introduction

A historic moment in the fight against climate change took place in April 2014. A coal plant in Ontario, the biggest province of Canada, burned what was promised to be Ontario's last load of coal ever. Thus, Ontario became the first large area in North America that quit coal burning altogether – a full year ahead of schedule. If Ontario (population 13.4 million) were an independent country, it would also be the first industrial country in the whole world to quit coal.

As recently as in 2003, a full quarter of Ontario's electricity came from burning coal. Now, a little more than a decade later, legislators in Ontario are planning to outlaw the burning of coal permanently. The change has been astonishingly fast, especially if one compares it to the oft-mentioned "champions of climate change" such as Germany and Denmark. These nations are planning to quit burning coal sometime between 2030 and 2050, assuming they can do it without significant economic costs and difficulties.

Ontario's achievement nevertheless remains almost invisible in the climate and energy debate. To our knowledge, not one environmental group has used it as an example, nor have they ever mentioned it as far as we know. Green politicians are not urging us to follow Ontario's example, even though Germany and Denmark are often highlighted as shining examples for all to follow. Yet Ontario's success is undeniable. We suspect the reason why it has been largely ignored is simple, yet troubling: Ontario produces more than half of its electricity with nuclear power, and has increased nuclear power production since 2003 by more than a quarter. The Canadians are winning the climate fight, but with what some call the wrong weapon.

Ontario is not the only example of nuclear power's abilities to reduce greenhouse gas (GHG) emissions. Two-thirds of all anthropogenic greenhouse gas emissions globally are due to energy production and use. Up to this point, nuclear power has been the undeniable champion in cleaning up energy production. When climate action is debated, this fact is more often than not totally ignored.

"Utter failure" is probably not too strong an expression to describe humanity's response to the climate crisis so far. At this point, snatching victory from the jaws of defeat requires us to use all the possible tools at our disposal. In the following pages, we will explain why trying to fight climate change with renewables and energy conservation alone seems like a foolhardy gamble. This is not merely our opinion: we shall demonstrate how even the forecasts made by Intergovernmental Panel on Climate Change (IPCC) largely agree with our thesis, even though details of these reports go unreported in mainstream environmental discourse. We will also explain how the ability of nuclear power to reduce emissions from energy production, and its comparable risks and downsides, have been at least to some extent, falsified and misreported. In the final part, we will offer some suggestions on what we can still do to mitigate dangerous climate change.

Both of the authors have been researching climate change and alternative energy sources for years. This book, an improved and modified version of the original that was published in Finnish in March 2015, is a distilled summary of some of our findings. Both of us are politically and economically independent but concerned citizens and energy researchers. Our main goal is to bring to the discussion viewpoints of which many people, even experts, have been largely unaware. We believe that effective energy and climate policy can be made only if the strengths and weaknesses of different solutions can be compared in a level-headed and evidence-based manner.

At certain points in this book, the reader could be excused for believing that we are hostile towards renewables, energy conservation or environmental groups. We wish to state most forcefully that this is absolutely not true. We think both renewables and energy conservation are essential tools in both the fight against climate change and our pursuit of a better and more just world. We also greatly value the work of the environmental organizations in many other fields, even if we have reason to believe that they are badly mistaken about the realities of both nuclear power and renewable energy.

We also believe that the widespread lack of critical thinking about renewable energy sources will end up harming the growth of renewable energy in the long run. These energy sources suffer from very real problems that are in most cases inherent in their very nature. Ignoring or downplaying these issues will make the problems only more difficult to understand and to solve. Lack of honest discussion will also erode peoples' trust in renewables as well as their proponents. The same goes for the environmental movement: their often dishonest anti-nuclear rhetoric, including but not limited to deliberate falsification of statistics (which we will discuss in more detail later) is already undermining their overall credibility. The solution to this is not for us or anyone else to be silent about such dishonesty. The solution is to stop being dishonest and focus on real solutions to tackling climate change while at the same time satisfying humanity's need for energy.

We do not hold any monopoly on the truth, and it is entirely possible that we are wrong about many things. We are fully prepared to change any of our views, claims and conclusions if we hear better and more logical arguments. This book already benefits from several corrections made to the original Finnish version: on the book's website, we have listed and corrected every mistake and misunderstanding that we or our readers have found from the text. While we hope that most of the errors have been caught by now, we are grateful for anyone who points out any flaws in our premises or our logic. We also host the discussion at our blog, climategamble. net, where we try to answer questions and discuss the all-important matters of energy, environment, climate change and a more just and sustainable world.

We owe thanks to numerous people who helped, knowingly or not, in the writing process. Special thanks go to Kaj and Jani for gathering interesting data. Thanks also go to Lauri, Maarit and Tanja for reading the draft and offering valuable suggestions. Lauri also provided invaluable help in matters such as publicity and getting in touch with interested and interesting people. Celia made a sterling effort in proofreading the English manuscript, and Tom offered valuable suggestions for this latest edition. Any mistakes left in the text are our own. Katariina Pekkola created the book's striking cover image, which we are very pleased with. Rauli would also like to thank The Finnish Association of Non-fiction Writers, WSOY literature trust fund and the Committee for Public Information in Finland for their monetary support in writing this book. Janne would like to offer his heartfelt thanks to his beloved Tanja, who helped to make this a book instead of a collection of separate articles.

Rauli Partanen   *Janne M. Korhonen*
Asikkala   Turku
                                                          Finland, 2015

The largest gamble of them all
We need everything we can get to quit fossil fuels

> **Key takeaways:**
>
> √ The whole energy mix, including electricity, heat (both space and industrial) and liquid fuels, needs to be mostly decarbonized in western countries by 2050.
>
> √ Even that might not be enough, and so far the climate fight has been an utter failure.
>
> √ There is very little chance of that happening unless we use absolutely all low-carbon tools at our disposal, to a much more aggressive extent than they are used today.

In order to have a chance at avoiding the potentially devastating effects of climate change, humanity must produce its energy almost entirely without emitting carbon dioxide by 2050. This requires much bigger investments than we are currently devoting to renewables, energy conservation and efficiency, nuclear power, and carbon capture and storage (CCS) combined. This is not an either/or question. We almost certainly need **all of the above**. In addition to cleaning up our electricity and liquid transport fuels, we need to drastically reduce the dependency on fossil fuels in industrial production, and at the same time turn the systematic deforestation of the planet into afforestation. These last two most likely increase the demand for carbon-free energy even further. Even if we succeed in every aspect detailed above, we are not going to halt climate change. But at least we will have a shot at keeping the risks of a climate catastrophe reasonably small. This is not our opinion, but rather the conclusion of the Intergovernmental Panel on Climate Change's (IPCC) latest (2014) reports, along with many other scientifically reviewed reports and studies. The clock is ticking, and the longer we wait, the deeper and faster emission cuts we will need.

If we fail to produce enough clean energy and cannot meet our targets in energy conservation, there is a real danger that we will simply end up burning all the remaining fossil fuel reserves as well. In the short run, these fuels offer tremendous economic benefits for their users. If there are not enough scalable options to replace fossil fuels, it might be impossible for the governments to ban or limit their use sufficiently, even if the political will to do so would be there. But if we burn even a significant portion of the remaining fossil fuel reserves, say more than a third, we are very likely to lock our planet in to a path to catastrophic climate change. What lies at the end of this path is anyone's guess, but it could well be the destruction of modern civilization, even humanity itself, and most of the Earth's current ecosystems.

Cleaning up our energy production might well be the humanity's defining challenge for this century. It is imperative that we transition to energy sources that emit low to no deleterious emissions in order to avoid drastic climate change. As the consequences of failing to do so are so dire, one would assume that all available tools would be at least considered. But the transition to clean energy has not been without controversy. At the moment, there are plenty of individuals and groups who have a strong interest in halting climate change, but are doing everything in their power to leave one of the most potent solutions – nuclear power – out of our very limited selection of available tools. Anti-nuclear activism has played an important role in the birth of many of our environmental organizations. Their continued opposition to nuclear technology seems to be one of the key reasons why these groups keep producing various reports, studies and scenarios that try to "prove" that the global energy problem can be solved by renewables and conservation alone. These reports can then be cited as "proof" that nuclear power is not needed, when someone – like us, when we first got interested in the subject – comes asking. Most people do not have the time, or the interest, to dig deep enough into the matter to see the numerous risks and pitfalls hidden in these reports. Indeed, it seems that most of the anti-nuclear activists themselves have missed or ignored these problems as well. Yet once one digs deep enough, as Google's now cancelled

RE<C project certainly did, one tends to find that renewable energy simply isn't enough[1]

Even more problematic than anti-nuclear activism alone is the fact that most environmental organizations oppose carbon capture and storage (CCS) technologies as well. This skepticism does have a slightly more solid foundation than the anti-nuclear arguments. Capturing carbon from the smokestacks of power plants and then transporting and storing it deep underground for millennia is actually very difficult. It has not been done on a very large scale anywhere. There is no guarantee that it will work as advertised and in a large enough scale. There are likely to be many unknown risks involved in sequestering huge amounts of carbon dioxide.

The problem is that by trying to ban both nuclear power and CCS, these organizations are trying to ban both the historically most effective way of cutting carbon emissions and what is potentially one of the most important future methods for emission reductions. In doing so, these organizations put the remaining options – mainly renewable energy and energy conservation – in a tough spot. Both renewables and energy conservation are very important tools, but current research strongly suggests that trusting only them to do the job is a gamble of the highest order. If renewables and energy conservation are the only tools that we are allowed to use, dangerous climate change can only be averted if everything goes perfectly according to the most optimistic plans laid out by renewables-only -advocates. There is simply no room for any unwelcome surprises, and zero margin for error. There is no backup plan. In addition, even the most optimistic renewable energy scenarios are likely to be too little, too late: they simply do not reduce emissions fast enough.

If these scenarios turn out to be overly optimistic, and nuclear power is opposed with the current unyielding fervor, the limitations of renewables will most likely be patched up with more fossil fuels. Even if the economic reserves of these fuels would turn out to be significantly more limited than commonly thought, we would still have enough of them to tip the planet into a climate catastrophe. In

---

1   See for example The Register's story on RE<C here: <tinyurl.com/nz5xqhd>.

addition, we simply do not have the time to run more experiments. It is remotely possible that renewables and energy conservation alone will be enough to stave off the climate crisis. The risk is that when we know beyond any possible doubt whether they work or not, it will almost certainly be too late to try anything else. Gathering evidence suggests it might already be too late. Nuclear power may be fiercely opposed, but at least actual history tells us it works. It can reduce emissions very rapidly and efficiently. Given all this, any climate policy that ignores some of the potential solutions is a gamble. The risks of such gamble must be compared with the risks and dangers involved in using nuclear power. We will discuss these risks later. A policy that seeks to ban outright the most important low-carbon energy source ever deployed is more than a gamble: it is reckless endangering of the future of our living planet.

Climate policy has been a failure

Despite all this, we do understand the point of view of many anti-nuclear people. Not too long ago both of us considered nuclear power to be at best risky and possibly entirely unnecessary technology, just as many still believe.

Our opinions changed slowly, as we both became aware of climate change and the grim realities of mitigating it fast enough. It was a rude shock to find that the world had made no progress at all in cleaning up its energy supply since 1990. Despite all the talks and all the negotiations, the world's energy mix is just as dirty as it was back when the Soviet Union still existed. At the same time, the total consumption of energy has grown enormously, so we emit more and more greenhouse gases every year. The solutions the public is left to believe will do the job – renewables and energy conservation – have not been able to even stop the growth of coal use, let alone reduce it.

When we dug deeper into the reports, statistics and studies, it became clear that the last time energy generation had been cleaned up with a speed and depth that matters had happened in the 1980s. Those same statistics also told us which technology had done it. This cast a totally new light on the rhetoric that was being fed to us. We had heard repeatedly that nuclear power played a marginal,

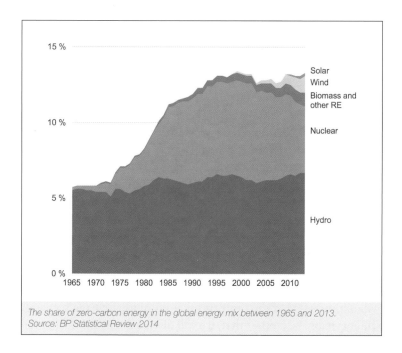

The share of zero-carbon energy in the global energy mix between 1965 and 2013.
Source: BP Statistical Review 2014

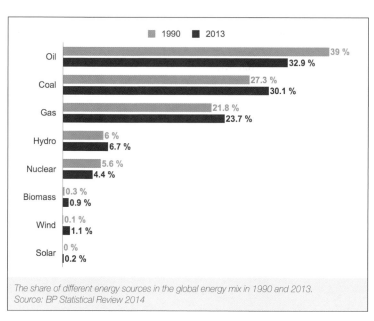

The share of different energy sources in the global energy mix in 1990 and 2013.
Source: BP Statistical Review 2014

even insignificant role in the global energy mix, as it produced only about 4.4 percent of our energy. But when the so called new renewables – mainly wind and solar – produce only about 1.5 percent taken together, were these politicians saying that these renewables combined played a role equal to about one third of "insignificant?" All of this resulted in a radical re-evaluation of our views. What else had we not been told?

To put it briefly: a lot. For many people, environmental non-governmental organizations are their principal sources of environmental education. While these organizations do not usually lie outright, they all too often misrepresent or withhold important information that would be required to form a balanced opinion and make informed decisions. This vital information is often scattered in small bits and pieces in the massive pool of scientific and technical literature. Balanced overviews that are both easy to read and comprehensive are hard to come by. To correct this deficiency, the next section will outline some of the problems that no-nuclear energy scenarios will inevitably face. After that, we will discuss the (mis)information that we often receive about nuclear power.

Studies are cherry-picked

Key takeaways:

√ According to IPCC estimates based on 164 scenarios (SRREN 2011), renewables (including all biomass and hydro) could, on average, produce roughly one third of our current energy use of 550 petajoules (PJ) by 2050.

√ Even Greenpeace's outlier scenarios could produce less than 80 percent of our current energy use by 2050.

√ By 2050, our energy use will likely grow by several hundred petajoules, as the developing world struggles to increase their economies, living standards and with them, energy use.

When the experts of the Intergovernmental Panel on Climate Change published their Special Report on Renewable Energy Sources and Climate Change Mitigation (SRREN) back in May 2011, several environmental organizations hurried to spread the news to the media: the IPCC is saying that a future based entirely on renewables is possible! Greenpeace's energy expert **Sven Teske** announced[2] that the report "shows overwhelming scientific evidence that renewable energy can also meet the growing demand of developing countries, where over two billion people lack access to basic energy services." The media around the world had a field day with optimistic headlines such as "IPCC says: 80 percent of the world's energy could be produced with renewables." These headlines are still referred to as proof on the redundancy of nuclear power.

Unfortunately, the report itself said nothing of the sort. It was written by hundreds of experts, had almost 800 pages, and presented 164 differing scenarios that reached all the way to 2050. These scenarios had two things in common. First, they all explored a future where energy would be produced only with renewables.

---

2   Greenpeace press release on IPCC SRREN report, 9th May 2011, <tinyurl.com/pgu8gtz>.

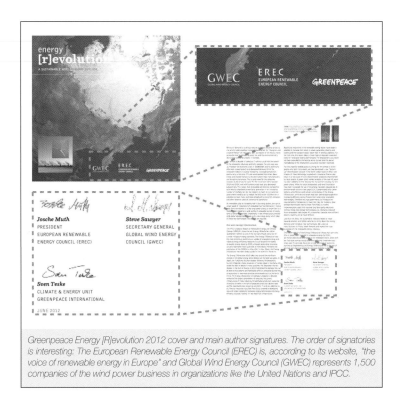

Greenpeace Energy [R]evolution 2012 cover and main author signatures. The order of signatories is interesting: The European Renewable Energy Council (EREC) is, according to its website, "the voice of renewable energy in Europe" and Global Wind Energy Council (GWEC) represents 1,500 companies of the wind power business in organizations like the United Nations and IPCC.

Second, they all ignored a variety of known and suspected problems of large-scale renewable energy generation. For example, small-scale firewood use for cooking was counted as a renewable and sustainable practice. No matter that it is documented to cause deforestation, erosion and millions of annual deaths – mainly women and children in poorer countries – due to indoor pollution.

Even with these rather generous definitions on what is renewable and sustainable, the report's conclusions were chilling. Every single one of the 164 scenarios showed that in 2050, renewables couldn't produce even the amount of energy we use today. On average, renewables would be able to produce about a third of the energy the world is likely to consume in 2050.

What on earth had happened? One explanation can be found from the misleading press release of the IPCC, which highlighted only some of the 164 scenarios in the report. One of these was ac-

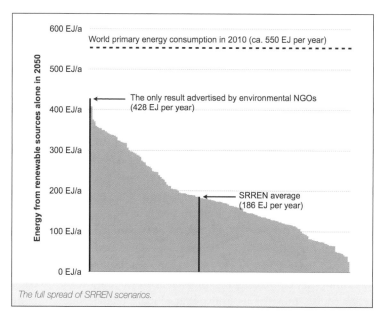

The full spread of SRREN scenarios.

tually a study commissioned by Greenpeace to explore the possibilities of renewable energy. This study was made in cooperation with organizations responsible for lobbying renewable energy in Europe and globally. In the 2012 version of Greenpeace's report, two of the three main authors of the text are actually employed by these lobbying organizations. Furthermore, Greenpeace's Sven Teske has openly admitted that key data required for the calculations came directly from the renewable industry.[3]

This scenario, by far the most optimistic in the SRREN, was a clear outlier. According to it, if the growth of energy consumption could be avoided almost completely, 77 percent of total energy used in the world in 2050 could be produced with renewables. Technically, the headlines were therefore correct. But the press release did not say anything about the fact that energy consumption would have to stay at near present levels. The 9 to 11 billion people living on planet Earth in 2050 would need to make do with about the same amount of energy that the seven billion of us use today, even

---

3   Sven Teske's blogpost on the website of Greenpeace International 16th June 2011, <tinyurl.com/ktzd73s>.

though even now billions live with poor or nonexistent energy access.[4] The press release also failed to mention that almost 99 percent of the scenarios in the SRREN report were much more pessimistic about renewables. Finally, nothing was said of the sad fact that even the most optimistic scenario would not be able to cut emissions fast and deeply enough.

Where does the blame go for such one-sided media coverage? Blaming the journalists for it is easy, but probably at least somewhat misguided in most cases. They simply do not have the time to read every press release critically, especially if the main message fits their preconceptions. If one wants to blame someone, one should blame those who wrote the press release and left the important conclusions unmentioned. If other IPCC reports were handled in similar way, reporting only the most optimistic cases and perhaps letting the fossil fuel industry write the highlighted studies, we would probably think that climate change in itself is only a marginal problem.

In reality, most of us know that climate change is one of the greatest challenges ever to face humanity, endangering ecosystems, biodiversity and possibly our society itself. The challenge is made even greater by the fact that this knowledge has been fiendishly difficult to channel into actual policies and action. What action, you may ask? Towards halting the use of practically all the fossil fuels during this century. Most of the work has to be done before 2050, and even after that, emissions need to keep falling. Eventually we need to achieve negative emissions, meaning that we must actively reduce the CO_2 concentration in the atmosphere. The sooner we manage this, the smaller the risk of runaway climate change that starts to feed on itself. As we write this, 2050 is only 35 years in the future. Any major changes in the energy system have historically taken at least half a century.

---

4   Around 5.6 billion of us have access to any modern energy services.

Is "wishing for the best" really the only plan?

Key takeaways:

√ The low energy -scenarios tend to forget the very real phenomenon called *rebound*, and often implicitly expect that the poor of the world agree to remain poor.

√ Scenarios where renewables alone can and will manage the decarbonization of our energy mix systematically dismiss the fact that their intermittency is a problem that will grow exponentially more complex as the share of intermittent generation in the energy mix grows larger.

√ To decarbonize our electricity, current renewables installation rates would need to grow by several orders of magnitude, and then stay at that level for the foreseeable future. This would represent build rates we have never witnessed even from all energy sources put together. To decarbonize the whole energy mix, that rate would need to be even faster.

√ The use of biomass, the cornerstone of many non-nuclear scenarios, would need to grow tremendously from current levels. Required increases would have severe effects for both biodiversity and food production.

√ Even with technological development, certain key minerals such as silver and tellurium are likely to become increasingly scarce before even half of our global electricity comes from renewables.

How much energy we can conserve?

To simplify a bit, we have two options for achieving the needed emissions savings in our energy system. Either we can increase the amount of clean energy produced by enough to meet total energy consumption, or we can decrease our need for energy so much that it does not really matter how we produce it. In reality, most agree that both options can and should to be pursued simultaneously. If we can decrease our energy use enough – be it through conservation or increased efficiency – we can meet this demand even if we do not use all low-carbon technologies at our disposal. But it needs to be remembered that with all the options available, we can meet the demand faster and more cost-effectively.

On paper, this seems like a reasonably easy problem to solve. The aforementioned scenarios from Greenpeace, along with corresponding reports and studies from World Wildlife Fund (WWF) and Friends of the Earth, rely heavily on energy conservation and a selection of their favorite low-carbon technologies to do the job. Their every scenario manages to stop climate change, erase the crushing energy poverty from the world, and even get the economy growing by building renewables and conserving energy. As mentioned above, these scenarios assume that 9 to 11 billion people living in 2050 will altogether use less energy than seven billion or so living today are using. According to many of these organizations, making this happen is merely a question of "political will."

However, any mention of something called the rebound effect (or Jevon's Paradox) is practically nonexistent in these reports.[5] Research on this phenomenon, well grounded in real-world evidence, predicts that when energy efficiency is improved, not all of the energy saved is actually going to be conserved. Instead, increased efficiency prompts people to use their efficient machines and appliances more often and for longer periods of time. For example, energy efficient LED lighting leads to increases in total use of light. More fuel-efficient motors and increased mileage per gallon lead to more driving, and so forth. It needs to be mentioned that efficiency improvements generally do save energy, and rebound only rarely turns to "backfire" where increased efficiency leads to increase in total

---

5   In both Greenpeace and WWF energy reports, "rebound" or "Jevons paradox" is mentioned precisely once. Greenpeace notes that energy use in India will grow rapidly partly because of the rebound effect, while WWF believes rebound effect can only threaten the electricity use reduction from more efficient home appliances. Both views are highly myopic and are very likely to prove grossly optimistic.

energy use. Nevertheless, rebound is still a largely unacknowledged problem. Even if rebound does not eat all of the savings we get by increased energy efficiency, it can eat a significant part of them, and wreak havoc with the more optimistic energy efficiency scenarios. Indeed, the most recent IPCC assessment report (AR5, 2014) notes – for the first time – that rebound effect may significantly limit the expected emission reduction gains from efficiency improvements. Today, IPCC admonishes future research to take rebound seriously instead of mostly ignoring it, as has been the case so far. A recent study found out that the rebound effect for British household gas, electricity and vehicle fuel use average at 41, 48 and 78 per cent, respectively.[6] What is puzzling is why it took so long for the energy researchers to acknowledge the rebound effect: after all, it had been described in the scientific literature already in 1865.

Demands to reduce our total energy use also raise questions about fairness and equality. To encourage people to conserve energy, it is often proposed that energy prices have to be increased. What is not discussed is the effects this will have on poorer people, how it will increase energy poverty, and what this will mean for many people who already face resource shortages and struggle to make ends meet. As an example, the fees German citizens pay in their energy bills to pay for the feed-in tariffs (FITs[7]) of renewables are actually a form of regressive taxation. This means that the poorest people suffer the most. Even as the German Energiewende is only starting, hundreds of thousands of households in Germany have had their electricity cut off because they've been unable to pay their bills.[8] Rising energy prices do not hurt the wealthier middle class as much, as energy represents a smaller share of their overall budget. In addition, they are the ones who can also invest in more expensive but also more energy efficient appliances, or even in solar

---

6   Chitnis, M., Sorrell, S. (2015), Living up to expectations: Estimating direct and indirect rebound effects for UK households, Energy Economics. doi:10.1016/j.eneco.2015.08.026

7   Feed in tariff is a price or a premium that a given energy production gets payed no matter what the actual price of energy is.

8   More on Der Spiegel 2013 article "Germany's Energy Poverty: How Electricity Became a Luxury Good." <tinyurl.com/ms5tvbq>.

photovoltaic panels subsidized by feed-in tariffs. From this perspective, the gushingly positive attitude usually displayed by the political left towards the Energiewende and its feed-in tariffs continues to baffle us: Larger wealth transfers from the poor to the wealthy will mean that much more poor households – possibly millions on the European level – will face increasing energy poverty in the future.

But it is on the global level where the "low-energy" scenarios that many environmental organizations are suggesting really seem to be a slap on the face for the world's poor. Greenpeace's Energy Revolution report plans that entire continent of Africa will use less energy than North America alone will use by 2050, despite the fact that Africa will most probably be home to more than five times as many people. This would leave the average African with less than ten percent of energy what average North American would have, and less than a third of what the average Chinese uses today. WWF makes similar assumptions in their report, giving those living without electricity less than one percent of the electricity that the average Finn uses today. No need to ask the poor themselves what they want: after all, they might give the wrong answers. A late 2015 Center for Global Development survey in Tanzania, for example, revealed that out of those who had experience with the kind of off-grid alternative energy environmental organizations are touting as the answer for the world's poor, full 90 percent wanted proper 24/7 electricity grid connection instead.[9]

It is possible that the poor of the world agree to remain poor and in the dark far into the future, but we would like to raise our objection on both the desirability and realism of these kinds of calculations. Even if we in the rich, developed world would want to limit the energy consumption of the poor in developing nations, we simply do not have any means to force them to do so. We think it is plausible, and for many reasons desirable, that in 2050 the average African would have access to at least the amount of energy that the average Chinese uses today. If this happens, the scenarios where energy demand stays essentially stagnant will most likely turn out to be incorrect.

---

9   http://www.cgdev.org/blog/dfid-solar-only-approach-rural-electricity-africans-want-on-grid Accessed 29.10.2015.

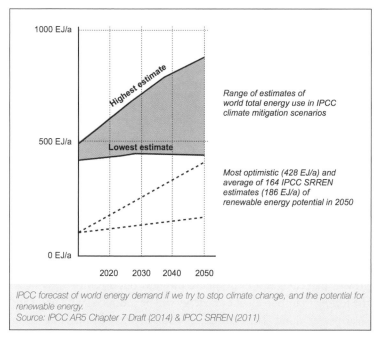

IPCC forecast of world energy demand if we try to stop climate change, and the potential for renewable energy.
Source: IPCC AR5 Chapter 7 Draft (2014) & IPCC SRREN (2011)

Because of their constantly increasing appetite for reliable energy, coal and other fossil fuels remain very attractive energy sources for developing countries in particular. This means that the emission trajectories of low energy growth scenarios might also be too optimistic. Since the development and wealth of the richer nations largely is and certainly was fueled by fossil energy sources, and given that we don't yet have examples of countries rising with renewable energy sources alone, we have only the flimsiest moral justification to try to stop the developing countries from following the same path. In addition, we lack the means to stop them even if we wanted to. This means that for both ethical and practical reasons, we in the rich countries have to reduce our own emissions even faster – with all the possible means we have at our disposal. Any coal burned in Germany simply so they could shut down their perfectly good nuclear power plants is coal that should not have been burned in the first place, but if it had to be burned, it should have been burned in a developing country to help raise their people from poverty.

To summarize, environmental groups assume that future energy demand will be exceptionally low. In most other forecasts, the world energy consumption will increase markedly by 2050, as can be seen in the graphic summarizing the IPCC's energy demand forecasts. These forecasts assume that we will fight climate change with much more vigor than we have done so far. If the current feeble excuses of proper climate policies prevail or if the developing world develops more than we've thought, energy demand in 2050 could well be much higher.

How long can renewables keep on growing?

The portrayal of renewable energy sources in popular media is misleadingly positive. The popular image of renewable energy makes it easy to imagine that renewables are a completely benign and problem-free way to produce energy. Optimistic articles gush about huge growth percentages and immense business opportunities, giving us the feeling that soon all of our energy will be produced with wind and solar power alone. These articles of course spread through the social media like wildfire. Amidst all of this hype and positive posturing, one question comes to mind. Why, then, do even the most optimistic renewable scenarios have very strict limits on how much energy we will use in the future? If renewables are set for exponential, endless growth, why do environmental organizations have to limit the future Africans to energy levels far below today's Chinese?

In their press release about the SRREN report, Greenpeace emphasized that even its very optimistic goal of 77 percent renewables by 2050 would only demand 2.5 percent of the "global renewable energy potential." If renewables have such a huge potential, why did every single one of the 164 scenarios in SRREN report conclude that in 2050, we will not be able to satisfy even our current energy demand with renewables alone? Had nobody told the top researchers behind IPCC's SRREN report about the advertised immense potential of renewable energy? Did they not know about it?

Of course they did. Scaling up renewable energy production unfortunately involves many difficult problems that have been largely ignored in the public discussion. To simplify, most of these problems are caused by the intermittency and variability of solar and wind power generation, while modern, industrialized world requires constant and reliable energy supply. When the share of variable energy sources in the total energy mix is small, these issues can be neglected. However, they will manifest themselves as the share of variable generation grows. This point is soon approaching in many European countries, and it will be reached far before renewables are a major source of total energy (not just electricity) use and sufficient emission cuts have been achieved. Currently it seems that adding more solar power in Germany will become very difficult before it reaches a share of ten percent of total *energy* consumption; in fact, growth in the rate of new capacity additions is already slowing down. Recently, not only Germany but also Spain and Japan have had to reign in the subsidy-driven growth of solar power, as the electricity grid is becoming more unstable and the subsidy (feed-in tariff) expenses are getting out of hand.

To solve the problems of intermittency, old thermal power plants have to remain operational. In other words, patching up unreliable renewable power generation in practice requires power plants that burn stuff, either fossil fuels, waste, or biomass. This is bad from the perspective of both climate change and added expense. Better alternatives, like energy storage and demand elasticity made possible by hypothesized smart grids, still need a lot of development and time before they are widely available. Greenpeace, for example, recognizes that these technologies will be essential, but it does not dare to give an estimate how much they will cost and when they will be available on a large enough scale. As a recent report[10] studying the decarbonization of the Finnish economy summarized the situation, "heavy reliance on intermittent energy sources like wind and solar requires a very rapid technological development and breakthroughs in energy storage, while it also includes many technical and economic uncertainties."

Local opposition to large, highly visible infrastructure projects is also slowing down the development of renewable energy production.

---

10 Lehtilä, A. et al. (2014) Low Carbon Finland 2050-platform. VTT 2014, <www2.vtt.fi/inf/pdf/technology/2014/T165.pdf>, page 4.

This can be seen for example in Britain, where new on-shore wind farms are becoming more and more difficult to build because of local resistance, and in Germany, where people are opposed to massive power grid expansions that would bring wind electricity from the windy north to densely populated southern parts of the country.

All this means that adding large amounts of renewable production may not be as easy as many would like to believe. Furthermore, wind turbines and solar panels have to be rebuilt eventually. Therefore, the current rate of capacity additions needs to grow by several orders of magnitude, and it will have to stay at that level indefinitely, if we want to have enough renewable production capacity to meet current and future electricity demand in a "sustainable" manner. If fossil fuels are to be replaced in other sectors of society as well, the rate of new renewable energy installation needs to grow several times more.

It would seem that renewables are now being asked to do something they might not be able to deliver. A recent (2015) study[11] examined eleven different emission reduction scenarios, all of which had received widespread media coverage. These included four scenarios that limited the energy toolkit to include only renewable energy. The authors of the study pointed out that all of these four scenarios assumed that energy efficiency would increase in the long term about twice as fast than its short-term record rate of increase has ever been. And this was only the beginning. According to their calculations, the building rate of renewable energy production capacity in the all-renewable scenarios would have to be 3 to 11 times faster than the best ever rate of new energy generation capacity the world has ever been able to achieve from all sources of energy production in total. The other seven scenarios, which did not limit themselves to only renewable energy, required build rates that were well within what is known to have been achievable.

In the light of these and other recent reports, recent trends in newly installed renewable capacity are more than worrying.

---

11 Loftus, P. J., Cohen, A. M., Long, J. C. S., & Jenkins, J. D. (2015). A critical review of global decarbonization scenarios: what do they tell us about feasibility? *Wiley Interdisciplinary Reviews: Climate Change*, 6(1), 93–112. doi:10.1002/wcc.324

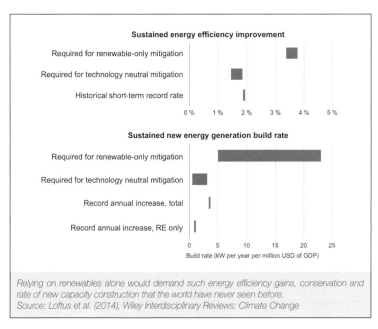

Relying on renewables alone would demand such energy efficiency gains, conservation and rate of new capacity construction that the world have never seen before.
Source: Loftus et al. (2014), Wiley Interdisciplinary Reviews: Climate Change

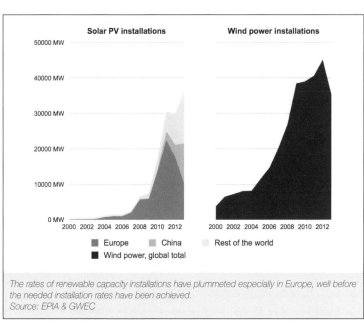

The rates of renewable capacity installations have plummeted especially in Europe, well before the needed installation rates have been achieved.
Source: EPIA & GWEC

Growth in annual additions to renewable capacity has been stagnating, even falling in some areas. Growth of wind power capacity has been slowing down, and in Europe, the rate of new solar photovoltaic (PV) installations has in fact fallen steeply. Even though significant quantities of both have been installed in recent years, the work is really just starting. The current rate of annual added capacity will not take us anywhere near what can even remotely be called a success. The installation trends of new solar PV capacity in Germany are particularly sobering. After peaking in 2011, they have practically halved every year for three years in a row[12].

If the current installation rate continues and the panels need to be replaced every 25 years (as is generally assumed), the German solar power production will have a maximum installed capacity of around 50 GW. This will suffice to produce less than 10 percent of Germany's annual electricity consumption, and far less of its total annual energy needs.

It seems that the reports and calculations that try to show us that nuclear power is not needed are reckless in at least two ways: They systematically underestimate the future energy demand, and they systemically overestimate the speed that renewable energy production is likely to grow in the longer term.

What about the ecological footprint of renewables?

At some point, adding new renewable capacity can hit an unexpected obstacle – environmental protection. Grand plans of renewable energy generation require even grander areas of land to be harnessed for energy production. Wind farms, solar PV farms, and the required transmission grids are only the beginning. All of these require massive mining operations to produce the raw materials required. Besides, enormous areas of fertile land need to be turned to industrial scale monocultures to grow biofuels. As an example, the entire Fukushima evacuation zone – of which most is about as radioactive as many residential areas in Finland[13] – could be covered in

---

12   http://tinyurl.com/phod8k3

13   Annual doses of 35 mSv per year are common in many residential areas in Finland. Roughly half of the evacuation area in Japan has an annual dose of

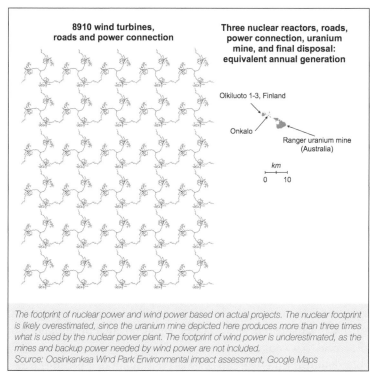

8910 wind turbines, roads and power connection	Three nuclear reactors, roads, power connection, uranium mine, and final disposal: equivalent annual generation

Olkiluoto 1-3, Finland

Onkalo

Ranger uranium mine (Australia)

km
0   10

The footprint of nuclear power and wind power based on actual projects. The nuclear footprint is likely overestimated, since the uranium mine depicted here produces more than three times what is used by the nuclear power plant. The footprint of wind power is underestimated, as the mines and backup power needed by wind power are not included.
Source: Oosinkankaa Wind Park Environmental impact assessment, Google Maps

solar PV panels but it would produce only about the same amount of energy that the destroyed nuclear power plant produced. It remains unclear whether massive additions of renewable capacity are compatible with demands to lessen human footprint on Earth and conserve biodiversity.

The problem is aggravated by the fact that most of the so-called "new renewable" energy generation today is actually that most traditional of all energy sources; burning biomass. This is in stark contrast with the popular image of renewable energy. While, for example, the pictures adorning WWF's energy report portray sleek wind farms and unobtrusive solar panels, the report actually tells us that even in 2050 a very significant part of global energy needs would be met by burning plant matter. Vast amounts of biomass are needed each year to satisfy the demand of these biomass-fired power plants

less than 20 mSv per year. http://tinyurl.com/othxwrt

and biodiesel refineries. In WWF's plans, by 2050 energy needs alone would require 30 percent more forest wood than is currently used for all purposes put together. In addition, around 250 million hectares of fertile land, an area comparable to the land area devoted to wheat today (240 million hectares) would need to be used to grow monoculture energy plants for energy production.

If we choose to heed WWF's advice and move towards the low-energy, even energy-poor future they envision, with only the tools they would allow us to use, we would need to find and develop previously unused fertile lands equaling the area that we currently use to grow our most important food plant, wheat. In addition, we would need to increase the human consumption of forest biomass significantly. All this needs to happen in a world where human population has grown by billions and agriculture is facing new challenges from worsening climate change. It should be noted that deforestation and agriculture are already the main causes for our current, alarming rate of biodiversity loss, causing what the researchers are calling the sixth great extinction. It boggles the mind that the organization proposing all this is actually one of the biggest environmental organizations in the world. No wonder, therefore, that WWF does not reveal where all these required lands will appear. With this in mind, it becomes easier to understand why in December 2014, 75 respected conservation ecologists and biologists co-signed an open letter to environmental organizations, pleading that these organizations would reconsider their view on nuclear power to help prevent the industrialization of the few remaining wildlife sanctuaries and to preserve biodiversity.[14]

Even if the required land areas would somehow appear out of thin air, it is far from clear that bioenergy can actually cut carbon emissions significantly. The more we invest in using bioenergy, the more problems this uncertainty could end up causing us. Yet biomass, usually disguised under the broader definition of "renewables", is often used to oppose a truly low-carbon option: nuclear power. In Finland, for example, the Green League's official 2014 alternative for a new nuclear power plant was a program of highly intensified biomass burning, with more or less cosmetic additions in wind and solar energy.

---

14   Read the letter here: <tinyurl.com/pc2wvow>.

The assumption that biomass is carbon neutral is largely a result of political expediency. When the IPCC originally considered the carbon footprint of different energy sources, very little research existed on the emissions from widespread biomass use. The emissions from coal and oil are relatively straightforward to calculate from the amount of fuel burned, but biomass presents a more complex case: fully accounting for their greenhouse gas emissions requires detailed knowledge of not just the type of biomass but also where and when it was harvested, and how the land use changed after biomass had been harvested. To keep things simple, and to the gratification of countries that used significant amounts of biomass-based fuels, the IPCC agreed that biomass would be counted as a carbon neutral fuel. This assumed that all the carbon emitted when biomass was burned would be sequestered back as the harvested plots grew back. Assuming a long enough time frame and no changes in land use, this is true. But two things need to be recognized. First, by the target year 2050, most biomass is not yet carbon neutral, and that is the relevant time frame on which we are operating. Second, biomass is only carbon neutral a few decades after we stop burning it. Before that, the continued burning of biomass for energy means that a big part of the carbon in our stock of actively used biomass is in the atmosphere at any given time, causing warming just like carbon dioxide from fossil fuel use. Currently, it seems that bioenergy does not satisfy even the European Union's criteria for renewable energy; the achieved emissions reduction compared to fossil fuels is too small and comes too late. If it is used to replace practically carbon-free nuclear energy, as just about every anti-nuclear organization and green political party worldwide is either openly suggesting or silently condoning, emissions can only increase.

Recent research suggests that biomass emission calculations are still somewhat uncertain. In favorable conditions, rapidly growing annual plants and logging residue such as wood chips can be relatively climate friendly. But under less favorable conditions, burning stem wood can be worse than burning coal. In theory, biomass can be used in a responsible, selective manner that minimizes environmental impacts and emissions. But if the demand for biomass keeps rising, the practice will most likely turn out to be something else.

The IPCC has increased its estimates of biomass emissions thirteen fold in just three years. Its current median estimate pegs biomass emissions in electricity generation to about half of those from fossil natural gas: slightly better than fossil fuels, but not enough to combat global warming. In addition, transporting and burning biomass releases particulate pollution that is harmful to human health and the environment.

When we broaden our view from electricity to the energy sector as a whole, we find ourselves staring at a gaping problem. Crude oil-derived liquid hydrocarbon fuels like gasoline and diesel are essential to keeping our society and economy running. Almost everything that moves runs on the internal combustion engine, which uses liquid fuels. Whether we want it or not, the choices made decades ago made sure that this will also be the case for many decades to come. We have built a world that runs on liquid fuels. It is slow and difficult to change to other power sources, such as electric vehicles[15] running on batteries. This is a very inconvenient fact, and it is often ignored, whether one looks at it from the perspective of climate change, or that of the myriad of problems caused by peak oil and tightening supply combined with growing demand.

Biomass is the only renewable energy source that we can turn into liquid fuels at prices even closely competitive with crude oil. If we use our entire sustainable biomass potential to generate just heat and electricity, will there be any left to replace even some part of our crude oil use? No. But replace it we must, for both climate's sake and for the sake of our energy security. It seems likely that the peak in crude oil production, together with rising demand in developing nations, makes sure that the oil market will likely tighten again sooner rather than later. We have already begun to substitute electricity for fuel in applications where batteries can compete with liquid fuels, and eventually we must also start to synthesize liquid

---

15 To give some sense of the scale of the challenge, a rough estimate about electric cars; with current technology, and assuming no competing uses for lithium or lithium-ion batteries, the world's annual lithium production would suffice for the batteries of about 3.6 percent of annual car production. Electrifying the world's 1.2 billion cars would also consume all the known lithium reserves on the planet, with none left for laptops, cell phones, etc.

fuels with electricity, but that requires abundant supplies of relatively cheap, clean electricity. Before that happens, we must increasingly rely on biofuels. This implies that we simply cannot burn the available biomass for electricity.

Even though the biomass problem is probably the biggest environmental elephant in the renewables parlor, it is not the only one. Other renewables beside biomass have their own, largely unstated environmental problems. For example, in order to increase wind and solar capacity significantly, we would need to increase the production of concrete, steel, copper and numerous other materials considerably. Even if material efficiency of these technologies is greatly improved, all this will mean more mines and more environmental damages in the form of open cast mines, tailings ponds, smelters and pollution. Somewhat ironically, the rare earths required in solar panels, electric cars and wind turbines even leave radioactive waste at scales comparable with uranium mining required for equivalent energy production. Even when uranium mining and the materials needed to build nuclear power plants are accounted for, renewables require on average at least as much mining operations as nuclear power, and often more. If all these are put together, renewables actually spoil much more land area than nuclear power – even if we include the areas polluted after nuclear accidents. While generalizations in this realm are difficult, it is entirely possible that health and environmental dangers are also far larger than from supposedly "dirty" uranium mining required for the similar energy generation. A further problem is that we are unlikely to even have enough of the certain critical materials for a renewable world. According to a recent study[16], transforming even 50 percent of global electricity production (not to mention total energy production) to renewables would require almost all of the known silver that exists in the world, and much more tellurium than is known to exist in known deposits.

---

16 Elshkaki, A., & Graedel, T. E. (2013). Dynamic analysis of the global metals flows and stocks in electricity generation technologies. *Journal of Cleaner Production*, 59, 260–273. doi:10.1016/j.jclepro.2013.07.003

Optimism is no guarantee of success

Key takeaways:

√ All major energy sources humanity has used so far have developed in a similar manner: with large growth percentages in the beginning and then leveling off. There is no reason to believe renewables would be an exception. Due to their intermittency, there is even reason to believe the contrary – that their growth will level off sooner.

√ Early growth of renewable energy has been fueled mainly by politically set tariffs – whenever these have been reduced, new renewable installations have plummeted.

√ The imminent breakthrough of solar and wind energy has been forecast for about a hundred years already.

In our experience, whenever someone voices doubts regarding the somewhat underwhelming progress that renewables have shown so far in solving the global energy problem, someone else can be found to push these doubts aside by saying that renewables are spreading at an exponential rate. If exponential growth continues, before long renewables will indeed be the only energy source we need. In theory, renewables have immense potential, and technological progress compared with decreasing costs will mean that more of them will be built – a lot more. The important question is threefold: first, do we know how long this exponential growth can last, second, will it be enough, and third, should we put all our eggs in this one basket?

Immense potential, at least on paper, and enormous growth percentages seem to make even many experts believe that the total victory for renewables is inevitable. It is good to remember that practically every energy source so far has grown in the exact same way: strong growth at first, which then levels off far before all of the energy in the world is produced by said energy source. For the mathematically minded, growth of energy sources has always broadly followed a logistic S-curve. In such a curve, slow start is followed by a period of exponential expansion. If one measures growth during this stage alone and uses very low starting point as reference, growth percentages can indeed seem staggering. But eventually the logistic curve will level off and rate of growth will drop. As far as we know, there is no indication or proof whatsoever that renewables would be an exception to rule. Indeed, the proof so far is in the opposite direction: as mentioned above, the rate of new solar electricity installations is already declining in forerunners such as Germany. Furthermore, intermittent renewables have severe, inherent problems that will make their growth much more difficult than the growth of more reliable, dispatchable sources of power. In fact, renewables have not even matched the growth rate of nuclear power in similar timeframe in the 1960s and 1970s.[17] None of this has prevented enthusiasts, or those who have monetary stakes in the game, from spreading hype how renewables will soon be our only source of energy. Below are just a few examples from history:

> *"We have proved the commercial profit of sun power [...] and have more particularly proved that [...] the human race can receive unlimited power from the rays of the sun."*

> *"Recent research suggests that a largely or wholly solar economy can be constructed in the United States with straightforward soft technologies that are now demonstrated and now economic or nearly economic."*

> *"Rapidly declining costs will make [solar, wind and geothermal energy] fully competitive in the near future."*

> *"The 100% renewable energy system [...] is no wishful thinking; it is a real policy option, in particular due to rapidly decreasing [renewable energy costs] and improving storage economics."*

---

17   That is, if we look at the actually generated energy. When measured in "installed capacity" or theoretical maximum generation, renewables have indeed grown faster than nuclear power. However, existing fossil fuel generation is displaced through energy generation, not with theoretical installed capacity.

The first quote is from the New York Times in 1916. The second was written by **Amory Lovins**, a well known nuclear critic, in 1976. The third is from a report published by the Worldwatch Institute in 1994. And the fourth is from an academic study published in 2014.

The current discourse shares some interesting similarities with the early discussion about nuclear energy. When nuclear power was at its infancy in the 1950s, it was treated much in the same way renewables are treated today. Before there was any widespread experience in actually building nuclear power systems and even less understanding of the associated difficulties, the popular press was awash with fanciful projections of world's problems being solved through the "unlimited power from the atom." Atomic energy was supposed to offer clean, cheap energy for all our needs, solve our environmental problems, lift the poor out of poverty, provide clean water for everyone, and even power our cars, trains and airplanes. As recently as in the 1970s, many experts asserted that most of our energy needs would be met with nuclear power by 2000. Only a few insignificant but surely easily and affordably solvable technical problems stood in the way.

With no more effort than some cutting and pasting, these claims could easily be recycled for the use of current "100 percent renewables" optimists. Their projections contain few if any problems that the "unlimited power from the rays of the sun" and wind cannot solve. What could also be recycled is the attitude towards people who dare to criticize the more optimistic visions. Both of us have repeatedly heard how we are old-fashioned dinosaurs who simply do not understand the overwhelming excellence and unstoppable progress of renewable energy and the technologies that go with it.

To conclude our message so far, we have explained some of the more significant, if under reported problems that plague intermittent renewable energy sources. The majority of respected climate and energy scientists at least tacitly agree that we need all of the available tools to solve the climate crisis. Plans to solve this crisis solely with renewable energy and energy conservation are based on overly optimistic assumptions about the success of both. Adding more renewables (by broad definition), will likely make them less than renewable, and could put "renewables" into increasing conflict with environmental and other values. Meanwhile the optimistic, uncritical handling of renewables in the media and the enthusiastic projections have lead even some experts to believe that the victory of renewables is inevitable. This is possible, but we simply have no proof, or even very good reasons to believe in the inevitability of a society based entirely on renewables. Instead, we have speculation that is based heavily on the trend lines of an industry that is just starting up. This kind of blinkered optimism, based on extrapolating early triumphs, has repeatedly proven unwarranted in the past.

We need to look at the facts objectively. We need to be able to change our views if studies and statistics give us reason for it. If we continue to deny or downplay the problems that increasing deployment of intermittent energy will cause, it will probably harm the prospects of these energy sources in the long run. In the same way, if the studies and statistics on nuclear disagree with our opinions and prejudices against it, we need to be prepared to change our views. Facts do not change because of opinions; opinions can, and should, change because of facts. To help with these considerations, the next part of this book seeks to present a discussion on nuclear power, and how it is generally treated in the news and media. We will also offer some examples on how even well-meaning organizations are trying to affect energy and environmental policy by distorting and even falsifying statistics.

From gamble to calculated risk

Lies, damned lies, statistics, and falsified statistics

Key takeaways:

√ Various environmental groups have been intentionally distorting statistics to make nuclear power look far worse climate mitigation tool than it really is.

√ To accomplish this, electricity from nuclear power is arbitrarily and with no scientific justification whatsoever assumed to emit as much greenhouse gases as electricity from gas or even coal plants.

√ Making policy based on these reports would reward a country that closed low-carbon nuclear and built instead basically anything from burning biomass to anything but the dirtiest coal, depending on the report in question.

√ Comparing even the badly failed Olkiluoto 3 project *alone* on a per capita basis to any national record speed of building solar and wind *together* will show nuclear power has been on the long term at least twice as fast in adding low-carbon energy generation.

Back in 2009, the World Wildlife Fund, one of the world's largest environmental organization, faced a problem. They were preparing a report[18] in which they would score industrial countries based on their success in mitigating climate change. The problem was that the wrong countries were winning. Since mitigating climate change relies heavily on low-emission energy production, the countries with the largest shares of low-carbon energy generation would lead the pack. Since nuclear power still holds the records for both speed and depth of emission reductions, the top positions were going to the countries that had decarbonized their electricity production with

---

18   WWF Climate Scorecard 2009, <tinyurl.com/oxg2tpm>.

nuclear power. Even more worryingly for the WWF's anti-nuclear message, these countries had done more for climate by accident during the 1980s than its chosen renewable energy champions had been able to do since 1990.

With a swift stroke of a pen (or perhaps an extra variable in the excel sheet), WWF decided to more than quadruple the electricity carbon footprint of these countries, namely France and Sweden[19]. The report was published, and the news headlines were predictable: Germany was the best of class in the fight against climate change! We are not aware of any mainstream journalist who has noted the small print hidden in the footnotes. It said that WWF does not consider nuclear power as a viable policy option and therefore counts it as having the same emissions as natural gas (350 gCO2/kWh). This was a completely arbitrary decision that finds absolutely no support from the mainstream energy and climate science. On the contrary, every thorough review of evidence from the likes of the IPCC, the United States National Renewable Energy Laboratory and many others has concluded that nuclear power is one of the cleanest energy generation options ever deployed. These studies show, for example, that nuclear electricity's greenhouse gas emissions are significantly lower than are those from solar photovoltaic electricity.

Such creativity with statistics is by no means an isolated incident, as revealed by The Climate Change Performance Index[20] published

---

19 German electricity production had a carbon balance of 495 gCO_2/kWh, Britain had 572 gCO_2/kWh and France had 362 gCO_2/kWh. In reality, and in all reputable global statistics and reports, France has a carbon balance of just 86 gCO_2/kWh, which is one of the lowest of all industrialized countries. This is explained by the large share of low-carbon nuclear energy in the French grid, and because of this, French citizens have very small carbon footprints. WWF clarified in its background material that since it does not want more nuclear, it chose to give nuclear power a larger carbon balance, similar to natural gas. This led to the quadrupling of France's electricity carbon balance from 86 to 362 gCO_2/kWh. Sweden, another country with low carbon electricity due to nuclear power and hydro, met a similar fate. Sweden has one of the cleanest electricity mixes at 47 gCO_2/kWh, and it was more than quadrupled to 212 gCO_2/kWh by WWF.

20 The Climate Change Performance Index, page 6. http://tinyurl.com/okefn2k

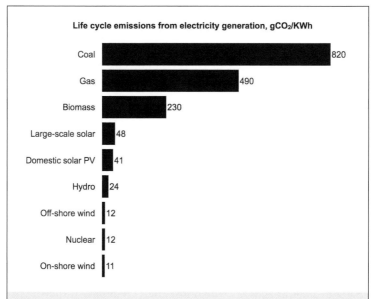

Life cycle emissions from electricity generation, gCO₂/KWh

Coal	820
Gas	490
Biomass	230
Large-scale solar	48
Domestic solar PV	41
Hydro	24
Off-shore wind	12
Nuclear	12
On-shore wind	11

The median carbon balance of different ways to produce electricity according to the IPCC (Life-Cycle Assessment, LCA). Nuclear power includes uranium mining, enrichment and fuel fabrication, plant construction, use, decommissioning and long-term waste management. Figures for wind and solar do not account for the necessary infrastructure for load following capacity or energy storage that is needed due to their intermittent production, so the chart is likely to underestimate their emissions.

by Germanwatch and Climate Action Network Europe in 2014. Again, the wrong countries were claiming the price, and again, the situation was fixed with creative carbon accounting for nuclear power. This particular index went even further than WWF did and declared nuclear electricity to have the same emissions as the dirtiest mainstream electricity, coal power. Given that this was supposedly a climate-focused index, it is interesting to note that a country could improve its score by replacing nearly emission-free nuclear with practically any mix of fossil fuels. One really cannot make this stuff up. We are sure that similar creative "indices" are already in preparation somewhere. Basing policy on deliberately falsified statistics and reports is of course a recipe for disaster.

We suspect that environmental organizations are in fact never going to tell us which countries have historically cut their carbon emissions the fastest and the most. The leaders in this game are those countries that built a lot of nuclear power in the 1980s, like France and Sweden. It is worth noting that these cuts were accomplished with technology from the 1970s, and were achieved completely by accident, as a by-product of energy policy enacted for completely different reasons. Despite even simultaneously increasing their energy use per capita, and without any climate policy at all, these countries got results that were significantly better and quicker to achieve than what Germany – or any country, for that matter – has managed with renewables and efficiency since the early 2000s. It is worth thinking what an active and evidence-based climate policy that pushed aggressively for renewables, energy efficiency and nuclear power could therefore achieve.

These climate reports and indices mentioned earlier raise one interesting question: If nuclear power is as useless in climate mitigation as environmental groups claim it is, then why there is a need to falsify statistics to make it look useless? Are these people afraid that anti-nuclear activists would lose their credibility if they report the actual science?

Speed is relative

Nuclear is often said to be too slow to build to make a difference for the climate. To support this argument, that poster red-headed stepchild of anti-nuclear movement, Finland's troubled Olkiluoto 3 project, is often trotted out as an example of how slow nuclear can be: an inexperienced contractor building a prototype reactor in a country with very stringent regulatory oversight has resulted in the reactor being nine years late and billions over budget. In Finland, the local Greenpeace chapter has habitually compared the build rate of renewables in Germany (population: 80 million) to build rate of Olkiluoto 3 in Finland (population: 5.5 million). Of course, such comparisons between two countries with 16-fold differences in population is fundamentally misleading. After we normalized the build rates with respect to population, we found that even the failed Olkiluoto 3 will actually be **almost two times faster** at increasing low-carbon electricity generation than the world record speeds of building both wind and solar power combined.

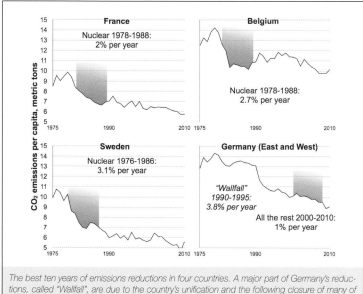

The best ten years of emissions reductions in four countries. A major part of Germany's reductions, called "Wallfall", are due to the country's unification and the following closure of many of ineffective power plants and industry in eastern Germany. In addition to these countries, also Belgium and Finland have cut their emissions markedly with nuclear power.
Source: CDIAC Carbon Data Project

Furthermore, when Olkiluoto 3 connects to the grid around 2018, it will produce more clean energy annually than all the wind turbines built in Denmark since 1990 produce today. When we look at national nuclear energy programs instead of single reactors, the record build rate of low-carbon energy production is more than five times faster than the equivalent record with "new" renewables (wind and solar combined[21]) so far. Even though we sincerely hope these renewable records will be broken in the future – as we have mentioned, the current record rates are far too slow to make much of a difference – it is clear that renewables still have a lot of catching up to do.

21 We exclude biomass and hydropower from this calculation, since significant increases in either run into hard limits of availability. Furthermore, both come with significant environmental price tags attached. Hence, lumping them with solar and wind when discussing renewables as tools for protecting the environment is highly misleading. While geothermal energy can be counted as truly renewable, so far (though with important local exceptions, as in Iceland and New Zealand) its contribution has been so minuscule that it can be ignored from the calculation without affecting the results.

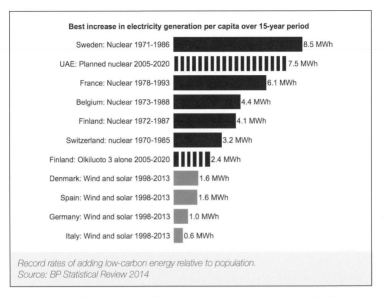

Best increase in electricity generation per capita over 15-year period

Sweden: Nuclear 1971-1986	8.5 MWh
UAE: Planned nuclear 2005-2020	7.5 MWh
France: Nuclear 1978-1993	6.1 MWh
Belgium: Nuclear 1973-1988	4.4 MWh
Finland: Nuclear 1972-1987	4.1 MWh
Switzerland: nuclear 1970-1985	3.2 MWh
Finland: Olkiluoto 3 alone 2005-2020	2.4 MWh
Denmark: Wind and solar 1998-2013	1.6 MWh
Spain: Wind and solar 1998-2013	1.6 MWh
Germany: Wind and solar 1998-2013	1.0 MWh
Italy: Wind and solar 1998-2013	0.6 MWh

Record rates of adding low-carbon energy relative to population.
Source: BP Statistical Review 2014

In light of real-life evidence, it would seem misguided to say that nuclear power is too slow to help significantly in the climate fight. Furthermore, if nuclear is too slow, doesn't this mean that renewables, particularly when considered individually, are utterly and hopelessly too slow?

When confronted with this question and the evidence above, some anti-nuclear advocates have moved on to conclude that even though nuclear power may be faster to build, it cannot reduce emissions enough as it only produces electricity, and a big part of our emissions come from other energy uses. Besides electricity generation, the most important sources of emissions in the energy sector are liquid fuels and domestic and industrial heat. Both are often expensive and difficult (but not impossible) to replace with electricity; this much is true. However, a telling fact about the state of energy journalism these days is that we are not aware of a single reporter who has asked the people opposing nuclear on this basis the obvious question: if more clean electricity cannot help in the climate fight, why on Earth are you advocating so strongly for solar panels and wind turbines?

In reality, of course, we simply have to figure out ways to supplant liquid fuels and domestic and industrial heat with electricity, or to produce the needed liquid fuels with electricity. Generally speaking, electricity is the only form of energy that is at least somewhat easy to generate with low enough emissions on a scale required to make a dent in the global emission curve. In all likelihood, ending our addiction to oil and other fossil fuels will require vast amounts of low-carbon electricity. Whether this electricity flows from turbines turned by steam heated from nuclear fission or from turbines turned by the wind does not really matter. What matters is that it is generated with as few greenhouse gas emissions as possible. In particular, liquid fuels and industrial heat are going to be difficult to replace. There is definitely too little public discussion on this matter. Only one thing seems clear: if we want to stop our oil and coal addictions in an environmentally sound manner, it will require enormous amounts of new clean energy.

What about the costs of opposing nuclear?

Key takeaways:

√ Many Green and environmental groups are prisoners of strongly anti-nuclear thinking prevalent during their founding in the 1970s and 1980s. Being locked in outdated mindset and in their own success in stoking anti-nuclear attitudes, these groups tend to oppose nuclear power first and worry about climate change and biodiversity loss later, provided as such worries do not threaten their anti-nuclear message.

√ Since bioenergy has been successfully framed as "clean and renewable" despite its very evident problems with both descriptions, the opposition of nuclear power often leads to more hacking, slashing and burning of our environment.

If one spends time campaigning against nuclear energy, that time, money and energy are not available for other purposes. For everything we do, we incur what is called an opportunity cost. In democracies, no group can successfully oppose all the options at the same time. If political capital and economic resources are spent to oppose nuclear power, there is less of them available for opposing power plants that emit greenhouse gases, for example. Perhaps the clearest example of this is bioenergy and how it is treated. Planting, harvesting and burning plants and trees for energy is potentially the most environmentally destructive way to produce energy. On one hand, environmental groups and the Green parties seem to realize this, but on the other hand, they realize that they need to present alternatives for nuclear energy. So while they may oppose industrial scale bioenergy in principle, they often also suggest increasing bioenergy use in order to replace nuclear power – or at least turn a blind eye to such plans[22]. The connection between opposing nuclear power and quite literally burning our forests instead is there, but many either do not see it or refuse to see it. On occasion this has led to Green parties being at least partly responsible for some of the most ecologically disastrous energy policy decisions in recent years.

Environmentalists cannot be blamed for the climate emergency; this should be self-evidently clear. But their chosen priorities, originating from a very different world of the 1970's and 1980's, make the climate crisis all that much more difficult to solve. Opposition of nuclear energy in all of its forms, from weapons to civilian nuclear power, has been one of the cornerstones of the modern environmental movement. As climate change rose to the list of our most important problems in the 1990s, environmental organizations had to react somehow. However, they have not been able to shake off the anti-nuclear sentiment that had become embedded in their organizational DNA, not to mention its success as a fund-raising tool. As a consequence, these organizations do an admirable job in spreading awareness about climate change and campaigning for radical emission reductions, but simultaneously use much of their energy and resources to oppose one of the most efficient ways to achieve those cuts. In Finland for example, the only issue that has

---

22   Good examples of this behavior can be found from the global proposals floated by WWF and more local proposals offered by the Finnish and Australian green parties, to pick only three examples.

caused the Green League to quit the government has been nuclear power. They have in fact exited the government over this matter twice. No environmental question, no matter how pressing, has ever received such a status even once. This tells a lot about the actual priorities of the party. As Lenin (who knew a thing or two about the difference between rhetoric and actions) is reported to have instructed, listen not to what the mouth is saying but watch what the hands are doing. The same is true for environmental organizations like Greenpeace. In our unscientific survey of media coverage of energy issues in Finland over a period of two years, for each press release condemning fossil fuels, Greenpeace Finland issued no less than six press releases condemning nuclear energy. For an organization that (rightly) argues that climate change is one of the most pressing issues of our age, Greenpeace seems to have its priorities exactly backwards.

Halting climate change is the goal, renewables and nuclear are tools

Key takeaways:

√ The goal and the tools are often mixed up for the detriment of serious discussion. The goal is to reduce emissions enough to mitigate dangerous climate change; the tools are, for example, methods for producing low carbon energy.

√ It is counterproductive to pit one tool against other tools, when the situation clearly demands the use of all the tools available.

√ There are different tools for different tasks, and a tool that works well in one place may not work as well elsewhere. For example, saving emissions with solar power in Germany can be ten times more expensive than doing it with wind power.

To be serious about stopping climate change, we need to learn the difference between goals and tools. Low-carbon energy generation, be it from renewables or nuclear power, and energy conservation are **tools**. The **goal** is, or should be, **to reduce emissions enough to mitigate dangerous climate change**. The goal is not to install solar photovoltaic panels, nor is it to conserve energy, nor is it to build more nuclear power, although all these can help us to reach our goal. In order to avoid dangerous climate change, we need to cut emissions as fast, efficiently, deeply and cost-effectively as possible.

Because we have spent so many pages exploring some of the weaknesses of renewable energy sources, one might think that we are somehow opposed to these energy sources. As we wrote in the introduction, this is not the case. Even though readability prevents us from stating this at every juncture, we wish to categorically state that we support (within environmental limits) all truly low-carbon energy sources and nearly all the energy efficiency proposals we've seen so far. Both of us have even invested our own hard-earned money in renewable energy generation. We will most certainly need a lot more renewable energy, and we are certain that a lot more is going to be built. The question whether or not renewable energy sources could play a major role in our energy system has already been answered: we know that they will be important sources of energy in the future. The only question is **how** important. We, of course, hope for the best and are excited about the potential and various benefits of renewable energy, and believe that solar and wind energy are amongst the most important tools for successful climate change mitigation.

But there are some things we are opposed to. We oppose how successes of some energy sources are weaponized as blunt rhetorical instruments for opposing other methods of low-carbon energy generation. While this sort of rhetoric is particularly noticeable when renewables are touted as the reason why nuclear shouldn't be built at all, some nuclear advocates also commit this error when arguing against renewables. We also oppose uncritical enthusiasm for renewables that currently accompanies much of the discussion, and prevents addressing their real problems in a timely fashion. Further, we oppose what we see as an unnecessary and harmful infighting between supporters of low-carbon energy sources, including the spreading of disinformation about "competing" energy sources. Finally, we oppose unwarranted optimism and breezy claims that the climate problem is easy to solve even if we leave out certain potential solutions. After all, more than 80 percent of the world's primary energy is still generated from fossil fuels: there is plenty work for everyone, and we most likely are going to need all the potential solutions that help us to reduce this percentage to as close to zero as soon as possible.

It is essential that we do not categorize energy sources as automatically good and desirable or automatically bad and undesirable. Instead, we need to see them as different, inherently imperfect solutions that have their pluses and minuses, and are suited for different tasks and different environments. For example, some renewables are better suited to some environments than others. To take just one example, one of the clearest lessons that can already be drawn from Germany's Energiewende is that installing heavily subsidized solar photovoltaic capacity in North Europe is currently not a very good idea. The massive solar capacity additions in Germany have resulted in some emission reductions, but these have been very costly. One study concluded that while the price of one saved ton of carbon dioxide emissions was 44 euros with wind power, it was 537 euros with solar PV[23]. Another study[24] revealed that Europe has wasted 86 billion euros (close to 100 billion USD) as renewables have been built in locations that are suboptimal or poor for them. In particular, the report questioned the wisdom of subsidizing solar PV in Northern Europe, and wind power in Spain.

---

23 Marcantonini, C. and Ellerman, A. D. (2013) The Cost of Abating CO_2 Emissions by Renewable Energy Incentives in Germany. MIT Center for Energy and Environmental Policy Research, CEEPR working paper 2013-005, <tinyurl.com/cglens5>.

24 World Economic Forum (2015). The Future of Electricity: Attracting investment to build tomorrow's electricity sector, <www.weforum.org/reports/future-electricity>.

Arguing against evidence: a primer

Key takeaways:

√ Anti-nuclear activists try to discredit the science and scientists supportive of nuclear power by accusations of corruption or global conspiracy.

√ Besides smear campaigns against science and scientists, alarmingly many anti-nuclear arguments are rhetorical and logical equivalents of argumentation used by climate deniers. While such arguments may contain a kernel of truth, they repeatedly leave out relevant details and comparisons.

√ Arguments labeling nuclear power "dangerous" rely on biased perception of risk where the mere possibility of some harm caused in the far future is given far more weight than the near-certainty of large harm caused today or in the near future.

The more we have studied the subject, the more we feel that many anti-nuclear advocates and climate change denialists share a common playbook. In both cases, much of the argumentation and rhetoric rests on carefully cherry-picked, generally non-peer reviewed "evidence," misleading or deliberately falsified statistics, and omissions of important information.

When writing this book, we asked for and received numerous studies and reports that anti-nuclear activists cited to back up their arguments. After careful reading and examination, we noticed that most of the studies were not peer reviewed, some were disturbingly flawed, and others did not actually say what the anti-nuclear activists were claiming they said. If we then countered the claims with peer-reviewed science and studies, our arguments were typically waved aside with accusations of "our" science being corrupt, or with quasi-philosophical arguments on the "nature of true knowledge" or the "changing nature of scientific knowledge." One gets the feeling that if we just wait and study a bit more, we will eventually find ultimate proof (which, of course, will not be subject to the afore-mentioned changing nature of scientific knowledge) that nuclear power is as bad and radiation is as dangerous as many anti-nuclear activists believe it to be, and suddenly millions, if not billions of people will die because of nuclear energy[25]. We will discuss the risks of energy production later in the book.

It is also all too common to move goal posts and change the rules of the debate whenever an argument becomes indefensible. When nuclear power is opposed due to its "inability" to reduce emissions (as it produces only electricity), the same argument suddenly ceases to exist if wind power or solar photovoltaics – both producing only electricity – are being talked about. If nuclear has the faintest possibility to kill any people at all in the future for example due to poor waste management (which we will also discuss later in more detail), it is seen as a valid reason to oppose all nuclear power. When coal and biomass burning kill seven million people every year, it is seen more or less as business as usual, as are deaths from oil and gas accidents. To loosely quote a tweet by Steven Darden: "Nuclear is dangerous because potential deaths are so much worse than actual deaths."

Anti-nuclear activists also have their pockets full of clever anec-dotes about the problems, damages and suffering somehow caused by the nuclear industry. Most of these are at least loosely based on actual events, and are therefore even more convincing, unless they are not challenged by anyone capable of putting these problems into perspective. As long as context and the problems of alternatives to nuclear power are ignored, one can easily get the impression that any reaction other than knee-jerk rejection of nuclear power can only result from either foolishness or from being a paid lobbyist.

In their book on the history of climate change denialism, *Merchants of Doubt* (2010), science historians **Naomi Oreskes** and **Erik M. Conway** wrote about the methods of creating confusion

---

25 For example Helen Caldicott, Chris Busby, Alexei Yablokov and many others have claimed this kind of death toll on Chernobyl and Fukushima accidents with very little to no scientific justification for their claims.

amongst the public despite clear scientific consensus. With example cases ranging from health hazards of smoking to ozone depletion, acid rain and climate change, they explained how the anti-scientific movement always operated with similar methods. Sometimes the anti-scientific movement began with honest skepticism, and the debate went on in scientific forums between people who actually knew what they were talking about. But even when the scientific evidence against the doubters became overwhelming, the issue rarely vanished: instead, true believers (and those with financial stakes in the game) shifted their strategy to keeping the "debate" alive in public forums populated by non-experts. This was done by spreading doubt and uncertainty about science, even years and years after the actual scientific debate on the broader matter had been settled. The media, always hunting for provocative headlines and "balanced" journalism, gave the anti-scientific crew time and space in no proportion to the actual scientific value of their message. This is particularly visible if the views and values of this minority happened to be mirrored by those of the journalist or the publication in question. Anyone who has followed the climate debate at all knows how the opinions of few climate deniers are given inordinate weight in conservative media outlets.

All too often, the anti-nuclear movement resorts to similar tactics. They do not have to challenge the peer reviewed science about the future of energy production, nor do they have to publish scientific papers on the effects of radiation. If your purpose is to spread fear, uncertainty and doubt, you do not have to be right and you do not need to offer much evidence to back your arguments. All you have to do is sow enough seeds of fear and doubt with scary images and clever anecdotes amongst non-experts. When you are helped by journalists who either share your opinions or profit from selling fear (or both), the task is not that difficult.

In fact, studying anti-nuclear rhetoric helps us to understand what motivates climate denialism. While it is undoubtedly true that some climate denialism (and some anti-nuclear advocacy) is fueled and funded by selfish interests, this cannot explain why so many well-meaning, intelligent people devote so much of their time arguing against mounting scientific evidence even without any material compensation. Rather, it seems that certain more or less factual questions tend to become questions of identity. Generally, climate denialism tends to take root within minds that stridently oppose any limits to "individual freedoms," and opposition to nuclear power seems to be something that largely defines self-described "environmentalists." In both camps, strongly held convictions and beliefs seem to be part of a tribal culture that separates "us" from "them," a friend from a foe. In such an environment, surrounded by like-minded people who share mostly information that confirms the group's pre-existing views, it is all too easy to fall victim to group-think, and to ignore countervailing evidence, while still retaining faith in one's own critical acuity. In fact, both climate denialists and anti-nuclear advocates often hold themselves to be **the** defenders of scientific method and evidence-based policymaking; both groups are very fond of citing researchers, facts and studies that seem to support their position; and both are quick to label contrarian voices as uncritical puppets of Big Climate/Big Nuclear (cross out as appropriate). We believe the vast majority in both groups are, generally speaking, sincere in their beliefs. However, being sincere in one's beliefs and being right are two very different things.

How dangerous is radiation actually?

Key takeaways:

√ Radiation is not as dangerous as it is commonly made out to be.

√ There are vast differences in radiation doses people receive from background radiation, but health records show little to no difference between these groups.

√ Average evacuees from Fukushima got a radiation dose that roughly matches the dose they would have received by living in Finland for a year.

√ In Fukushima, fear, anxiety and social stigma along with forced and prolonged evacuation cause much more harm than the actual radiation would.

√ Nuclear power is statistically one of the safest, if not the safest, way to produce energy – even after harmful effects of uranium mining, accidents and nuclear waste are accounted for.

One of the most popular residential neighborhoods in Finland is located on a scenic Pyynikinharju ridge a mere stone's throw from the center of Tampere, Finland's third largest city. Straddling the two lakes of Tampere, this moraine relic from the last Ice Age is dotted with traditional, picturesque wooden houses, lovingly cared for by the well-off residents who have been able to afford a house and yard mere kilometers from the city center. In recent years, house prices in the area have increased by as much as 20 percent, and the trend seems set to continue. More and more people want to move to this beautiful, accessible yet quiet neighborhood. They do not seem to be deterred in the slightest by the fact that in places, the area is significantly more radioactive than the average in the infamous ghost town of Pripyat, next door to Chernobyl.

The bedrock of our planet contains significant quantities of radioactive elements, including uranium. When uranium inevitably decays, it releases an odorless, invisible and radioactive noble gas called radon. In places where bedrock is close to surface, and in particular in places where suitable natural formations exist to collect and concentrate the seeping gas before it disperses into the atmosphere, radon can be encountered in very high concentrations. Throughout Finland, with its landscape scraped nearly to bedrock by the glaciers of the last Ice Age and crisscrossed with radon-collecting moraine ridges, hundreds of thousands of people receive annual radiation doses that are significantly higher than what the great majority of the Japanese evacuated from Fukushima received.

Exposure to radon and to its decay products gives the average Finn an annual radiation dose of about 1.6 millisieverts[26] (1.6 mSv). This is about half of the total *background radiation* dose, 3.2 mSv, received by Finns on average every year. The rest of the dose is mostly comprised of other natural sources of radiation and medical X-ray examinations. The average is only slightly above the global average background radiation dose of 3.01 mSv. However, according to estimates made by Finnish radiation protection authorities, of Finland's approximately 5.5 million inhabitants at least a hundred thousand people receive annual doses ranging between 10 and 20 millisieverts from purely natural sources. Doses ranging to 35 millisieverts per year or more are not unheard of[27]. To put this in perspective, the "liquidators" drafted to clean up the Chernobyl disaster received on average a single dose of 150 millisieverts. Survivors from Hiroshima received a single dose of about 200 mSv.

---

26 Sievert and millisievert (shortened Sv and mSv) are units of radiation exposure, used to determine the health effects of radiation on humans. Due to space constraints, radiation biology in this book is somewhat simplified; however, for a very readable account on the intricacies of radiation and its effects, we heartily recommend the book "Radiation: What Is It, What You Need To Know" by Robert Peter Gale and Eric Lax (2013).

27 The highest annual averages measured in Finnish houses are 30,000 Bq/ m^3 (translates roughly to 600 mSv/y), and momentary concentrations have been as high as 100,000 Bq/m^3 (which would equal, if continual, a dose of about 2,000 mSv/y). The highest measurements taken from the soil have been around 1,000,000 Bq/m^3.

From Fukushima, the single worst exposure to general public is believed to be no more than 50 mSv, while one third of the evacuees received between one and five millisieverts, and approximately two thirds received no more than one millisievert. Because the average background radiation dose in Japan is only about two millisieverts per year, this means that two thirds of the evacuees received total doses equal to, or less than, what they would have received had they lived in Finland for a year. We hope no one mentions this to the thousands of Japanese tourists flocking to Finland every year. Importantly, these doses were one time only whereas background radiation doses are continuous. A 15-year old Finn from high background radiation area has therefore already received more radiation than the average liquidator at Chernobyl.

Finland is by no means unique. The world is dotted with regions with high natural background radiation. Some are located in areas where radioactive minerals are abundant. Others can be found from mountainous areas where there is less atmosphere to block cosmic radiation. Locales with exceptionally high background radiation can be found in Iran, Brazil, India, Australia, United States, and China. Record-setting dose rates are found from Ramsar, Iran, where doses may exceed 200 mSv per year. However, Finland presents a special case because of its comprehensive health services and meticulous record keeping. This, along with other similar records from other countries, the comprehensive health survey of Hiroshima survivors, and a century of study on the effects of radiation on humans, allows us to draw some conclusions on the effects of radiation on human health.

What we know is this: while very high doses of radiation are undoubtedly harmful, radiation is by no means as dangerous as it is generally believed to be. For example, since Finland in general and its high-radiation areas in particular do not experience inordinately high numbers of cancers or other illnesses, it is very likely that the unfortunate evacuees from Fukushima will not do so either. What illnesses may occur are very unlikely to be caused by radiation. Scientists generally agree that at doses below 100 millisieverts, any health effects are most likely to be too small to be detectable in any statistical survey of population.

Why, then, are we concerned about radiation at all? The key reasons are that we know that very high doses are harmful, and that gathering solid evidence about the effects of low doses has been exceedingly difficult. This has led the world's leading scientific bodies to start from what we do know: high doses are dangerous. Prudent and cautions scientist as they are, in absence of solid evidence they have to assume for the sake of caution that risks may be directly related to the dose, even down to very low levels.

The danger from high doses is fairly well established. We have evidence that if twenty people receive a dose of 1,000 mSv each, we should expect that approximately one of them would, at some point in their life, develop a cancer because of this radiation. If the dose is increased to 2,000 mSv, there would be on average two extra cancers. However, what is more controversial is whether the relationship holds linearly as the dose goes down. As of now, the majority of radiation protection scientists agree that for precautionary reasons, until we have conclusive evidence to the contrary, we should assume that this is the case. For purposes of calculating risk, these experts assume that if two hundred people are exposed to a dose of 100 mSv each, we should expect one of them to develop a cancer, and if two thousand people are exposed to 10 mSv each, again one case of cancer should result. This assumption is known as Linear No-Threshold model (LNT). As its name implies, it assumes that the long term health damage caused by radiation is directly proportional to the dose received, and that there is no "threshold" below which the dose has no effect. Furthermore, it is often assumed that only the total dose matters. Under such an assumption, a single dose of 100 millisieverts has similar effects to ten years of living in an area where background dose is 10 millisieverts per year.

We already know that, strictly speaking, the LNT model cannot be the whole truth. A single dose of about 8,000 millisieverts or more is considered to be irrevocably lethal even with the best treatment. However, many radiation therapies produce nearly equivalent doses but spread them over days and weeks. The largest total dose a human has been known to survive is whopping 64,000 millisieverts, received by a man named Albert Stevens ("patient CAL-1") over a span of 21 years because of a reckless plutonium injection experiment in 1945.[28] Stevens lived for over two decades with significant quantities of plutonium embedded in his body, emitting radiation all the time, and eventually died of unrelated heart disease at the age of 79. Yet no human is known to have lived more than a day or two after rapidly receiving doses in the region of 30,000 millisieverts or more. According to the basic LNT model – the source of the claim "there is no safe level of radiation" – either Stevens should have died no later than days after his total dose exceeded 8,000 millisieverts, or people who received one-time doses upwards of 30,000 millisieverts should be alive and kicking. It is therefore plausible to theorize that human cells have mechanisms that allow themselves to repair at least some damage caused by smaller radiation doses. Evidence for such mechanisms has indeed been found in laboratory experiments.

Nevertheless, questions remain, and finding answers has proven to be difficult. Because 25 to 40 percent of people in industrialized countries develop cancer anyway, ferreting out the effects of small radiation doses amongst so many confounding factors has proven to be a formidable challenge. The effects are simply so small that they are lost among the statistical noise, as there are things out there which are much more carcinogenic than modest radiation doses. Red meat, for instance, may well be one.

Solid evidence of harm begins to crop up once radiation dose exceeds 100 mSv, but below that, uncertainty reigns. Recently, several research programs have been initiated, and it is possible that the next ten years will shed some more light into matter. So far, experiments have suggested the existence of cell repair mechanisms, as mentioned above. But on the other hand, some large-scale meta-studies have uncovered indications of small but statistically significant health effects appearing at doses well below 100 mSv. Furthermore, a small but vocal minority of radiation researchers still believes that small doses of radiation are actually beneficial to health. According to this so-called radiation hormesis hypothesis, small doses stimulate the body's natural defense mechanisms and therefore help fight diseases.

---

28 Welcome, E. (1999). The Plutonium Files: America's Secret Medical Experiments in the Cold War. New York; Dial Press.

So far, most reviews of relevant science have concluded that evidence from humans does not support the hypothesis, even though some evidence from other species and from laboratory experiments would seem to do so. The existence and continuing support for the hormesis hypothesis among some radiation safety scientists should be kept in mind whenever fringe theories from the other end of the spectrum are trotted into the limelight. Although there are some researchers who argue that small radiation doses are far more dangerous than radiation biology generally assumes, on balance there seems to be more evidence and more researchers favoring the hormesis hypothesis. Although it is often believed to be somewhat of a fringe theory, the continuing allure of hormesis hypothesis is in itself an important reminder of how relatively harmless radiation is. Even if evidence does not support the hormesis hypothesis, scarcity of solid evidence of harm at low doses (directly caused by the fact that small doses are not very dangerous) also means that all the evidence cannot conclusively refute it.

While we may not know precisely how harmful small radiation doses are, we have solid evidence to believe that they are far from being the imminent danger to health and safety that some anti-nuclear campaigners would wish us to believe. Even if studies show conclusively that small doses do have some health effects, it is certain that the effects cannot be very large. In other words, small doses cause at best a small risk. If it were otherwise, there would be no controversy over the hormesis hypothesis, and national cancer registries in countries with high background radiation rates would have already sounded the alarm.

However, there are two things that we know with reasonable certainty. One: the dangers of radiation are, generally speaking, greatly exaggerated. Two: in practice, this fear-mongering causes more problems and even medical issues than radiation does.

Fear mongering is more dangerous than radiation

One of the most surprising facts about nuclear accidents is that they are actually not very dangerous. Even (and perhaps especially) when nuclear accidents are compared with other energy-related accidents, health damage and death toll from nuclear energy remains very small. However, the mere possibility of nuclear accidents arouses visceral fear. Talk of nuclear accidents evokes images of deserted wastelands where scarcely a weed can grow, of mutations, and of silent dangers for centuries to come. One of the commenters on the early manuscript of this book captured this sentiment perfectly, noting that whenever someone mentions a nuclear accident, the first images that pop into mind are desolation at Hiroshima and the napalm-burned children of the Vietnam War. Such images are familiar to many of us, as are the desolate radiation wastelands and radiation mutants pictured in science fiction books, movies and video games.

Radiation seems to scare people at some deep, fundamental level. We cannot see, hear, taste, feel, or smell it, but we know that it may harm us. However, radiation is very easy to detect with relatively primitive instruments. In medicine, radioactive elements are used as tracers precisely because they are so easy to detect. At best, we can detect even the decay of single atoms. Therefore only very small amounts of tracers need to be inserted into our bodies, keeping harm and discomfort at a minimum.

We believe it is precisely this property of radioactive substances – that they are so very easy to detect – that is the main reason why nuclear accidents cause such widespread fear, panic, and media frenzy. In time, even minuscule leaks of radioactivity will be detected, sometimes from the other side of the world, and click-baiting media can proclaim with the boldest type: "RADIOACTIVITY DETECTED!" However, the amount of media coverage nuclear accidents receive bears no relationship whatsoever with how dangerous, statistically speaking, radiation or nuclear accidents actually are.

As far as accidents go, nuclear mishaps tend to unfold at a decidedly leisurely pace. Even if significant quantities of radioactive materials escape, as happened in Fukushima, people usually have plenty of time to move away from contaminated areas before receiving really dangerous doses. Chiefly for this reason, both the World Health Organization (WHO) and UNSCEAR (United Nations Scientific Committee on the Effects of Atomic Radiation) estimate that radiation released from Fukushima will most likely have no statistically observable health effects anywhere in the world, ever[29]. Even amongst emergency workers, the health effects are unlikely to be noticeable. This, remember, was an accident that resulted in three uncontrolled reactor meltdowns.

The worst civilian nuclear disaster ever, the infamous Chernobyl accident in 1986, is likely to cause noticeable health effects. Even so, the best estimate[30] of a hundred or so leading radiation scientists representing eight international organizations and the governments of Belarus, Russia and Ukraine is that these will number 4,000 at most. Most of them will be thyroid cancers. Even these would have been almost entirely preventable had the Soviet government ordered a speedy evacuation, forbidden the use of contaminated vegetables and milk, or at least issued iodine pills to the exposed population. Any one of these measures taken alone would have greatly limited the dangerous uptake of radioactive iodine-131; taken together, they would have been very effective in reducing the incidence of cancer. However, no such measures were taken until it was far too late. So far, the confirmed death toll from Chernobyl's radiation stands at 15 deaths from thyroid cancer, and 28 emergency workers who received massive doses of radiation during their attempts to extinguish the burning reactor, and died of acute radiation poisoning or its aftereffects. Even though rumors abound, credible evidence of widespread harm beyond these cases is scant or nonexistent.

Nevertheless, for the sake of discussion, it is instructive to take at face value the reports of stringently anti-nuclear organizations. According to Greenpeace[31], for example, 93,000 people have died

---

29   United Nations Scientific Committee on the Effects of Atomic Radiation, 2013. UNSCEAR 2013 Report, Volume I: Report to the General Assembly. Scientific Annex A: Levels and effects of radiation exposure due to the nuclear accident after the 2011 great East-Japan earthquake and tsunami, <tinyurl.com/pkmgbq4>.

30   The Chernobyl Forum: 2003-2005 (2006). Chernobyl's Legacy: Health, Environmental and Socio-Economic Impacts, <www.iaea.org/sites/default/files/chernobyl.pdf>. Represented organizations are IAEA, WHO, UNDP, FAO, UNEP, UN-OCHA, UNSCEAR, and World Bank, alongside Belarus, the Russian Federation, and Ukraine.

31   Greenpeace (2006) The Chernobyl Catastrophe: Consequences on Human Health, <tinyurl.com/4oyn978>.

or will die because of the disaster. This number is reached by, in short, adding up all the increased mortality since 1986 in areas that received any fallout. One of the many peculiarities of this method is that this figure includes diseases such as cirrhosis of the liver. There is absolutely no evidence whatsoever of radiation causing cirrhosis; on the other hand, there is ample evidence that frequent use of one relatively commonly ingested chemical will do so. After 1986, the fallout area has seen an economic and social dislocation incomparable in recent history, and partly as a result, alcohol use in the area, as in other areas of former Soviet Union, has skyrocketed. This confounding factor did not inconvenience Greenpeace's body counters (who insinuated that because exposure to radiation can reduce resistance to infections, cirrhosis cases – which may be also caused by viruses such as hepatitis – must be caused by radiation), but it should give pause to anyone interested in basic scientific integrity.

The direct health impacts of radiation due to nuclear accidents are very likely to be relatively small. On the other hand, psychological impacts such as stress, anxiety and self-destructive behavior are known to cause severe problems. The aforementioned report on the health impacts of the Chernobyl accident 20 years after the event found out that the psychological effects are most likely the biggest health threat both for those still living in the area, as well as for those evacuated from it. Psychological problems manifest themselves as increases in the prevalence of suicides, abortions, alcoholism, drug abuse, depression and a general increase in risky behavior, which is believed to be caused by the "survivors'" fear of radiation and the misconception that they are doomed because of it. No matter that these individuals would likely receive similar and possibly greater radiation doses if they moved to the scenic neighborhood near Pyynikinharju, Finland.

Similar problems have already been reported among Fukushima evacuees, and no doubt many of them will live in fear for the rest of their lives. By the most recent estimate, 1,600 people are believed to have died because of the evacuation. Many of these were elderly or infirm whose illnesses were so serious that they were unable to cope with the disruption, but perfectly healthy people have been driven to suicide by fear and stress. Numerous others have undoubtedly suffered lasting psychological damage.

The lesson is clear: societies need to be prepared for rare but possible disasters. Action plans, including plans for disaster communication, need to be prepared in advance. The public also needs to be educated realistically about the dangers of nuclear accidents. Neither the government nor the nuclear industry can be allowed to lapse into complacency, much less arrogance: when nuclear power is going to be used, accidents and mishaps will happen, although with the new generation of safer reactors these will be even rarer and their impact will be even smaller. Attempts to assuage the public with claims of "perfect" safety are counterproductive; however, spreading knowledge about the likely effects and how to take protective measures is not.

But there is an elephant in the room that polite conversationalists tend to ignore. What responsibility, if any, should the anti-nuclear movement and its fear-mongering participants shoulder for the damages caused by nuclear accidents? For decades, the anti-nuclear movement has been very adept at stoking people's fears with all sorts of claims and innuendo only rarely supported by scientific evidence. Moreover, those who oppose nuclear energy have made practically no attempts at all to distance themselves from the worst of the tinfoil-hat brigade and the most blatant of charlatans. There seems to be no claim so outlandish about the dangers of nuclear energy that it warrants a rebuttal from even those organizations, like Greenpeace, who claim to base their opinions on science.

Based on extensive research in Chernobyl and on the first assessments of the effects of Fukushima, we now have good reason to believe that the primary cause of health and other damages from a civilian nuclear accident – even of the very worst kind – is fear. It is fear of radiation and its effects that cause people to worry, stress, flee, and refuse (or be prohibited by the government) to return. It should go without saying that radiation should be treated with respect, and unnecessary exposure, especially to high doses that we know to be harmful, should be minimized. Still, we argue that nuclear accidents would be far less harmful were it not for the decades-long scare campaign that many well-meaning but misguided anti-nuclear advocates have been running. Should they bear some

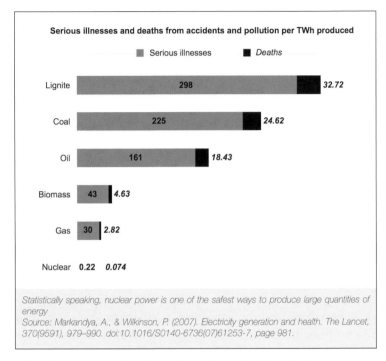

Statistically speaking, nuclear power is one of the safest ways to produce large quantities of energy
Source: Markandya, A., & Wilkinson, P. (2007). Electricity generation and health. The Lancet, 370(9591), 979–990. doi:10.1016/S0140-6736(07)61253-7, page 981.

of the blame? Does freedom of speech include the freedom to cause great harm, even if indirectly, by spreading unwarranted fear and stress, to those already suffering a crisis?

And what of the media and its responsibility? The Fukushima nuclear accident, with zero casualties, was the main headline throughout the world for the whole spring of 2011. The Western media in particular almost ignored that a massive tsunami following an earthquake under the Pacific Ocean had just wiped out entire towns, killed 16,000 people and forced hundreds of thousands more to leave their homes. Fear sells, and any mention of nuclear sells, so nuclear fear sells even more. A vicious circle ensues: The more nuclear fear sells, the more fear is stoked, and more there is incentive for media and for publicity-dependent organizations to capitalize on that fear. The reporters and fear-mongers may be simply unscrupulous click-baiters and quacks, or they may be well-meaning but misguided; the results are the same. In an environment like this, how can level-headed assessments compete?

Risk-free energy isn't

No matter what the reason, a premature death is always an unqualified tragedy. So is the loss of a home or a livelihood due to a disaster, be it a nuclear one or something else. But one thing needs to be made clear: No energy source is without risk or harm.

When a record "once in 2000 years" rainfall in 1975 broke the Banqiao dam in China, around 26,000 people died almost instantly, 145,000 others perished in the aftermath and 11,000,000 people lost their homes[32]. Even in well-off Europe, there have been dam failures that have claimed hundreds or thousands of lives[33] more or less instantly, with little warning or possibility for evacuation. The normal operation of coal mines and coal power plants cause millions of premature deaths every year. Even in Europe, these deaths are in the tens of thousands every year, with the United States suffering around 7,500 deaths annually due to particulate pollution caused by coal burning[34]. The most pessimistic peer-reviewed analysis on the health impact of Fukushima pegs the nuclear damage worldwide during the next 40 years to be less than what can be expected from particulate pollution caused by one large coal plant in one year of normal operations[35]. This, mind you, is before the damages and risks of vast CO_2 emissions are even considered. No wonder that respected British journalist and environmental activist George Monbiot noted[36] in the aftermath of Fukushima that "while nuclear causes calamities when it goes wrong, coal causes calamities when it goes right, and coal goes right a lot more often than nuclear goes wrong."

In reality, nuclear power causes much less harm than the burning of coal or biomass and the pollution they release. Even Greenpeace's

---

32   More information: <en.wikipedia.org/wiki/Banqiao_Dam>.

33   For a list of dam accidents, see <en.wikipedia.org/wiki/Dam_failure>.

34   See <www.catf.us/fossil/problems/power_plants/>.

35   Ten Hoeve, J., Jacobson, M., (2012), Worldwide health effects of the Fukushima Daiichi nuclear accident, <tinyurl.com/k2l7e55>, DOI: 10.1039/c2ee22019a.

36   Monbiot, G. (2011) Japan nuclear crisis should not carry weight in atomic energy debate. The Guardian, 16.3.2011, <tinyurl.com/o6lspd9>.

own reports conclude as much. In different publications, Greenpeace has raised alarm for health effects of coal and nuclear power. Comparing the aforementioned Greenpeace report on Chernobyl with their 2013 report[37] on coal power's health hazards, we noted that if the cost of replacing just 300 of the largest European coal plants with nuclear would be a Chernobyl-scale accident every ten years or so, it would represent a significant improvement in terms of public health – even if we entirely forget about the carbon dioxide released by the coal plants. Besides particulate pollution, coal plants are the major reason for elevated levels of mercury and many other heavy metals in the environment: for example, the consumption limits placed on many species of Baltic fish are mainly due to coal burning.

It comes as a surprise for many that coal plants in fact release even far more radiation to the surrounding environment than nuclear plants do in their normal operation. Coal contains radioactive elements like radium, thorium and uranium. Even though their concentrations are small, the massive quantities of coal burned every day throughout the world mean that vast amounts of this waste escapes even the most effective anti-pollution systems, spreading to the environment. To be clear, these radioactive elements do not pose a significant health risk compared to other toxins like mercury and other heavy metals and particulate matter released by coal plants, which kill by the thousands every day. The difference is that nuclear plants collect and store almost all of their waste, but coal plants as a rule do not. Even if their massive piles of coal ash are disposed of "safely," they freely use the atmosphere as a garbage dump for their carbon dioxide and at least some of the ash. Furthermore, even the "collected" ash is very often simply dumped in huge toxic waste ponds, whose failures have repeatedly inundated large areas with a sludge pregnant with poisons like heavy metals and dioxins.

The significant distinction that may explain the difference in our attention to damages actually caused by coal plants, and those largely hypothetical damages that nuclear power could cause, is

---

37 Greenpeace (2013) Silent Killers: Why Europe must replace coal power with green energy, <tinyurl.com/ohjj366>.

that even small radiation releases are easy to detect and measure. Furthermore, nuclear power plants are required to report their mishaps. Easy detectability combined with the fear factor associated with radiation and nuclear power means that any nuclear accident, no matter how minor, will be in the headlines. Actual deaths due to burning coal or biomass usually appear only in local health statistics, without gathering much interest in the media or amongst the public. Even the slight possibility of a health risk due to radiation is much more newsworthy than thousands of actual deaths caused by the coal industry each year. This results to the same "availability heuristic" that causes us to be afraid of terrorists even though actual risk of being harmed in a terrorist attack is far smaller than the risk of being struck by lightning.

Similarly, although the tens of thousands of sudden deaths caused by hydropower dam failures have not been widely publicized, it does not mean that dam failures do not exist or cannot happen ever again. And while wind and solar power seem to be fairly risk-free (if one turns a blind eye to the mines where their materials come from), they still cause their share of industrial accidents. When these accidents are tallied, it is not entirely clear whether these new renewables really are significantly safer than nuclear power. For example, a research briefing[38] commissioned by Friends of the Earth UK from the University of Manchester's respected Tyndall Centre studied the relevant scientific literature and concluded:

> *"Overall the safety risks associated with nuclear power appear to be more in line with lifecycle impacts from renewable energy technologies, and significantly lower than for coal and natural gas per MWh of supplied energy."*

However, the main risk of relying on renewables alone is the enormous risk of failing to address climate change fast enough.

---

38   Tyndall Centre, University of Manchester. A Review of Research Relevant to New Build Nuclear Power Plants in the UK. Including new estimates of the CO2 implications of gas generating capacity as an alternative. A research briefing commissioned by Friends of the Earth England, Wales and Northern Ireland, 2013. Page 16, <tinyurl.com/oz6b47o>.

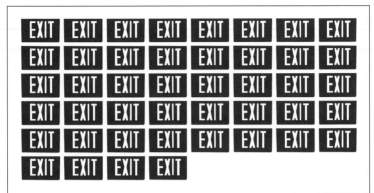

One of the news about Fukushima that went viral was that 40 trillion Becquerels of tritium was leaking to the Pacific Ocean. This amount is equal to about 44 self-luminescent EXIT signs, of the type that has been widely used for example in the United States. None of the news reports dealing with the issue ever put the leak into context.

Whatever the risks of a major ramp-up in nuclear power, the risk of insufficient action on climate change is greater by several orders of magnitude. This is the core argument of this book, reflected even in the title.

In our experience, anti-nuclear argumentation has quite rapidly shifted away from safety issues after Fukushima. This is particularly noticeable among the more educated nuclear naysayers. Based on interviews and discussions we have conducted or participated in, we have reason to suspect that most of the active and more knowledge-able anti-nuclear activists are perfectly aware that modern nuclear is about as safe as any energy source, and that dealing with nuclear waste is not the big and unsolvable problem it is often claimed to be (more on the topic later). Many readily admit so, and claim that their current opposition is based on economic issues alone. Then again, these people seem to have no problem having regular people as well as rank and file anti-nuclear activists believe that nuclear power is an exceptionally dangerous source of energy.

The risks and disadvantages need to be compared fairly

Key takeaways:

√ Mainstream media and environmental organizations tend to use vague and/or opaque units and terminology when reporting about radiation, and almost never provide relevant comparisons. Such reporting tells non-experts almost nothing about the true risks involved.

√ There is reason to suspect such vagueness is a direct result of the inconvenient truth: almost always, radiation releases reported are not nearly as harmful as the anti-nuclear activists would like them to be. But vague language allows the reader's imagination to fill in the blanks with desired horrors.

In reports and presentations supplied by the anti-nuclear movement, the risks and dangers of nuclear waste disposal, nuclear power generation or uranium mining are almost never measured in any meaningful way. They are even more rarely compared with any other relevant risks and dangers – such as those posed by the most likely alternatives for nuclear power. This does not mean that anti-nuclear organizations give no information on radiation and health hazards. It is the style and framing – the implications – of this information that is important. For example, these reports typically measure radiation in Becquerels (Bq). It is a unit that sounds foreign and dangerous; furthermore, it is a very small unit whose use results in reports with many zeroes behind every number. The catch is that Becquerel itself tells us absolutely nothing about the health hazards of a given radiation source[39].

---

39 Becquerel (Bq) is a unit of how many atoms disintegrate (and therefore give radiation) in a second. The natural radioactive isotopes in a human body disintegrate at a rate of about 4,000 Bq. A banana has isotopes enough for about 15 Bq, and so forth. To know anything about the actual danger or health hazard, we must know at the minimum what the isotope is, how it will

When reading reports by anti-nuclear organizations, it is a rare thing to find any meaningful estimates of actual health hazards. For example, we have never seen any estimates what problems, a leaking nuclear fuel repository could cause for the people in the area. It is of course impossible to exactly predict the future, but we can make educated guesses. We can prepare scenarios and forecasts that give us some idea of various ways of how things are likely to unfold, and what the effects will then be for the environment and us humans. Scientists do this routinely in climate science, which is in many respects a much more complex matter. And scientists have done it with nuclear power[40]. These concrete, often publicly funded and readily available peer-reviewed studies almost never make it to the aforementioned reports, however. Why? We fear the reason is that their results do not support the anti-nuclear agenda.

Instead, many if not most anti-nuclear organizations seem to prefer relying on vague claims and (to lay people) obscure units of measurement when discussing the dangers of nuclear power. This is a fantastic tactic for spreading fear, uncertainty and doubt. Without actually making outrageous and easily refuted claims about atomic destruction and radioactive wastelands, these organizations can enlist people's imaginations to fill in the gaps. One is left wondering whether this actually is their purpose, or whether this is merely a strategy that has been found to work wonders.

---

behave in the human body (e.g. whether it will be taken up by some specific organ, and how long it will last in the body), is it an internal or external exposure, and if internal, how the isotope got inside the body.

40   The EU-funded ExternE –study is one example of a large study that accounted for accidents, nuclear waste and other hazards, and still found nuclear power to be one of the safest way to produce energy. Image 12 a few pages earlier, estimating health damages from various energy sources, relies heavily on ExternE data.

How serious is the nuclear waste problem?

Key takeaways:

√ Storing nuclear waste, or spent fuel (High Level Waste or HLW), is more a political problem than a technical one. We have several technically sound and economically sensible solutions for the waste, from deep geological repositories to burning the waste in fast reactors.

√ The long-term risks from nuclear waste (for example, a repository leaking) are rarely quantified and never compared with any other relevant risks. There are grounds to suspect the reason is the fact that these risks are so small that most comparisons would immediately show how meaninglessly small the risks of waste disposal are likely to be.

√ Living atop a badly leaking geological repository would give a person a radiation dose roughly equal to eating a bunch of bananas every year – in the worst case scenario.

√ The nuclear industry is one of the few industries that is required to collect and store its wastes in the first place.

In our experience – nuclear waste, or more precisely, spent nuclear fuel – is one of the main reasons many people oppose nuclear power. It is therefore instructive to use this as an example of what the scientific method actually reveals about the matter. The key questions are whether nuclear waste can be (or even needs to be) stored "safely" for a very long time, and what the word "safely" actually means in this context.

Very few people are familiar with what nuclear waste actually is, and how it could possibly harm the health of humans or nature. Most people get their nuclear waste education from mass media, including comics such as The Simpsons; there, nuclear waste is treated routinely as some utterly horrifying and mysterious substance, endowed with almost magical powers for harm (occasional superhero origin stories notwithstanding). This being the case, the ground is ripe for fear-mongers. One only needs to mention the words "nuclear waste," and without needing to expend any effort on details such as elucidating the mechanisms through which nuclear waste could escape its confinements or cause harm to people, the listeners are all too ready to fill in the blanks with terrifying yet unspecified dangers. The studies and evidence that we actually do have on the matter suggest something completely different; even in worst case scenarios, the harm caused by a nuclear waste repository going horribly wrong will be very small and limited to a very limited area very close the repository.

Nuclear power is the only major energy industry, and one of the few industries in general, that is, broadly speaking, required to collect and store all of its waste. The safe final disposal of high-level nuclear waste – namely spent nuclear fuel – has therefore been studied for decades. Even if political maneuvering and NIMBY politics[41] have hindered the progress of building deep underground storage facilities, for example, there are several viable solutions to choose from. The safety of these geological storage options has been studied extensively. One of the central aspects of these studies is modeling how the storage could leak and what harmful effects such a leak could have for the environment and people living nearby.

One of the most advanced final disposal projects is Onkalo in Finland. Posiva, the company building the disposal site, has studied its safety aspects for years, and its conclusions have been reviewed and vetted by external experts. One of the studies[42] systematically

---

41 NIMBY refers to Not In My BackYard, a common problem when anything significant needs to be built. Its more severe form is BANANA – Build Absolutely Nothing Anywhere Near Anything. In Finland, the NIMBY aspect and local opposition to a disposal facility was largely overcome by an open decision-making process, which ultimately resulted in two municipalities competing for the opportunity to host the facility. The outcome is starkly different from e.g. the government-dictated process in the United States, which ultimately contributed to the abandonment of the Yucca Mountain project.

42 The text refers to scenario PD-BC on page 137 in the report Hjerpe, T.,

calculates the radiation doses that future people could be exposed to in different scenarios if the repository would fail and leak. The results are very interesting and eye-opening. The most pessimistic outlier scenario could result in the most highly exposed person receiving an annual radiation dose equivalent to eating a bunch of bananas.[43]

Although the details differ, the Finnish case serves as a good example of how spent nuclear fuel may be disposed of in the future. The radioactive elements within fuel are kept separated from the environment by multiple, independent "release barriers." These include the ceramic fuel itself, which is quite resistant to corrosion; a thick copper canister, where several spent fuel bundles are stored within a cast iron framework; a bentonite clay buffer around the canister and as backfilling in disposal tunnels; and, finally, some 400 meters of granite bedrock. The aforementioned calculation assumed that the copper canister is damaged already when it is put into the storage. In addition, both the canister and the bentonite clay surrounding it are expected to vanish mysteriously in just 1,000 years. In reality, the canister is assumed to last at least tens of thousands of years, and there are bentonite clay deposits around the world that remain intact hundreds of millions of years. Either one of these barriers alone would be enough to stop the leaks almost completely. Moreover, just to be on the safe side, the calculation assumed that the person getting the dose lives his whole life on the most contaminated square meter of soil, eats only food grown on that square meter, and only drinks water that is also gathered from that square meter. If any of these assumptions is tweaked to be even a bit closer to reality, especially if the person eats something else as well, the amount of "banana doses" drops significantly from a bunch of bananas to a dose of less than one banana. To make it clear

---

Ikonen, A. T. K., Broed, R. (2010). Biosphere Assessment Report 2009. Posiva 2010-03, <www.posiva.fi/files/1230/POSIVA_2010-03web.pdf>. Another, more recent paper with even smaller doses predicted is Posiva (2013), Safety Case for the Disposal of Spent Nuclear Fuel at Olkiluoto – Biosphere Assessment 2012. Posiva 2012-10, <www.posiva.fi/files/3195/Posiva_2012-10.pdf>.

43 More specifically, about 0.00018 millisieverts per year.

once more, this is the absolute worst-case scenario in the study. All the rest result in markedly smaller doses.

It is, of course, Posiva's calculation, albeit reviewed by external experts. It is healthy to be skeptical about studies made by corporations that have their own stake in the game[44]. Such calculations should be criticized, and they have been. For example, Greenpeace has commissioned their own review on the matter. At least part of this critique deserves attention: for example, regulators need to ensure that tight schedules do not result in Posiva cutting corners when it comes to safety. Despite everything, even Greenpeace's critique nevertheless has to conclude that "…there is no proof so far that the planned repository is not safe and that the open problems cannot be solved…"[45]. Of course, not all problems can be foreseen. However, the worrying *effects* of the most pressing potential problems according to Greenpeace – such as the copper canister and bentonite clay giving way well before they are supposed to – are already factored into Posiva's worst-case calculations. Even so, the outcomes for the poor fellow living in the worst contaminated spot while growing all of his or her food at the same location remain far less than what even the most cautious observers would suggest is needed to cause any noticeable harm.

What is left unsaid?

When digging deeper and deeper into the nuclear waste debate, we noticed an interesting rhetorical trend. For all the verbiage expended on attacking nuclear power and nuclear waste disposal, the most interesting things were those that were left unsaid. For example, in the nuclear waste case, while entire web pages suggest various ways in which the disposal could possibly fail, not a single one ever made an effort to estimate what the practical consequences of such failure might be. There was no rigorous (or even less than rigorous) estimation even closely comparable to Posiva's models; no calcu-

---

44 This, of course, should apply also to the renewable utopias discussed earlier.

45 Lempinen, A. and Silvan-Lempinen, M. (2011) Reverse Logic – Safety of Spent Nuclear Fuel Disposal. Greenpeace International. Page 38, <tinyurl. com/pmkbqvf>.

lation that showed even noticeable harm to people possibly living close to a nuclear waste disposal site; nor were there any estimates of how large of an area a leaking disposal site might contaminate – and how badly. The closest that the anti-nuclear rhetoric gets to such estimates have been various sound bites about how seemingly small quantities of plutonium could, in principle, pollute entire lakes were plutonium to be distributed evenly into lake water – although left unsaid is how that dispersion is to be achieved, much less sustained over time[46]. We can only conclude that the lack of such estimates reflects a significant problem undermining much of the anti-nuclear rhetoric: the results of more rigorous studies simply do not support many of the implicit allegations.

The nuclear energy debate in general differs from almost any other debate in the absoluteness of the demands made. We have been told repeatedly that there is, for example, no way to guarantee that a nuclear waste repository will never leak. Similarly, we hear claims that there is no way to guarantee that nuclear accidents will not happen ever again. Technically speaking, these are of course correct claims: no one in their right mind should ever say that an accident could *never* happen, nor that a repository *cannot* leak under any circumstances. The repositories are designed for tens or hundreds of thousands of years: those are long timespans by any measure, and very weird things may happen. However, the crucial question is evaded: "so what?" What if there is a leak? How large a leak? What will happen to the people or the environment? Is it going to be a serious problem, and what is serious anyway? How could the events unfold and how likely are they?

Anyone arguing against nuclear energy on the basis that perfect guarantees of safety are impossible to come by is actually subscribing to a magical or homeopathic view of radiation. In this view, radioactive substances and "artificial" radiation are absolute evils

---

46 Another thing left unsaid was that by same measure and under similar conditions, caffeine could be even more deadly than plutonium. In 1976, a U.S. professor of physics, **Bernard Cohen**, offered on public television to eat the same quantity of plutonium oxide (the most common form finely distributed plutonium dust takes) as anyone would eat pure caffeine. There have been no takers so far, and for a good reason.

that simply cannot be tolerated at all. No matter that there really isn't a meaningful, effects-based distinction between "natural" and "artificial" radiation, the holders of this view see any "artificial" radiation sources as fundamentally intolerable no matter what the dose or actual effects. This view is implicit in numerous articles and arguments that raise alarm whenever "radioactivity" or "radioactive substances" are ever detected anywhere or there is a possibility of their release, regardless of the amounts detected or postulated.

However, if we take these arguments to their logical conclusion, we will quickly note that any human activity could be condemned on exactly the same grounds. There is absolutely no way to guarantee that any action or inaction will not cause serious harm to people or the environment in a hundred thousand years. Similarly, if we apply the homeopathic view of harmful substances more widely, any activity potentially producing any amount of anything harmful could be condemned. What of, say, wood fires? After all, carcinogenic soot from burning wood spreads easily thousands of kilometers. We believe that most anti-nuclear activists, even those who use the absolute guarantee argument, are reasonable people who are perfectly aware of the absurdity of these latter claims. However, it does not seem to prevent them from applying one standard when arguing about nuclear energy, and another when arguing about something else. If one were truly interested about increasing knowledge and looking for solutions, one should be asking questions such as "how much?", "so what?", and "what about risks of the alternatives?" It should be remembered that one of the alternatives in this case includes the scenario where we fail to address climate change.

Nothing in the above means that the calculations, models and estimates we have quoted here (or other similar studies) are infallible truths. There are always uncertainties, not only in models and in their calculations, but also in the fundamental assumptions about – for example – what radiation does to a human body. New scientific knowledge can overturn old assumptions, and it is perfectly possible that we may turn out to be in the wrong. However, this does not mean that the models and estimates are useless. Even if they are probably not exactly correct and may turn out to be little more than guesswork, they are far better guesses than those achieved by intuition alone. We are the first to concede that engineers and scientists make errors, can be overconfident, and very likely have a bias to produce reports that satisfy their employers. Nevertheless, in the Posiva example above, the margin of error remains comfortingly large: even if the result is off by a factor of one hundred thousand, the dose will still be so small as to pose, at worst, a negligible risk.

Estimates such as the ones above are never even meant to be the final word in any matter. Rather, they are ways of reducing the associated uncertainty. By putting one's assumptions into numbers, one advances the conversation about risks and benefits in a way that those relying on adjectives alone can never equal. Therefore, it is more than a shame that anti-nuclear organizations generally do not engage the models and their outcomes directly, and try to supplant them with better models. Of course, one possible explanation could be that they simply do not have anything credible to offer, and have to resort to mere rhetoric.

Nuclear waste repositories are not a new thing

An oft-heard comment is that we cannot know what will happen to nuclear waste repositories in geological timescales. In the strict sense, this is true. What is left unsaid here is that we, however, have some experience on what may happen to nuclear waste. Nature has been burying it since back when life consisted of single-celled organisms only.

We were somewhat surprised to learn that no less than sixteen natural nuclear reactors have been discovered from Oklo and Bangombé in Gabon back in the 1970s. These sites aroused intense scientific interest, and it was quickly concluded that the reactors came to being about two billion years ago when ground water began to seep through an extraordinarily rich uranium deposit. This water sufficed to moderate a chain reaction that kept the reactors "operational" for perhaps half a million years or so, albeit at a low power compared to power reactors of today. However, the reactions and their end results were exactly the same as those occurring daily at nuclear plants of today. Therefore, the waste products were also identical, and geologists were able to trace their movements in the ground. Despite the fact that the reactors and their waste were for the most part close to the surface and in contact with flowing streams of ground water for unimaginable stretches of time, most of the dangerous waste had traveled less than a few meters from its point of origin. No matter the shapes of continents were meanwhile transformed beyond recognition and life itself had time to evolve to all its current splendor.

Across the globe, other so-called "natural analogies" to nuclear waste disposal can also be found. Under Cigar Lake in Canada lies a very rich deposit of uranium oxide, the same stuff that, in its enriched form, is used as fuel in most nuclear reactors today. More than a billion years old, this deposit was found by chance because a naturally formed layer of relatively porous clay (less effective than the bentonite used in nuclear repositories) completely blocked any tell-tale elements from escaping. Numerous ice ages have rolled over the deposit, with very little to no effect. Other analogues can be found from places as varied as Loch Ness in Scotland and Sicily.

Because of these analogues, we know that storing the waste relatively safely for hundreds of thousands of years is within the realm of possibility. After all, nature has managed this repeatedly by pure accident. As discussed above, we also have good reason to believe that even relatively serious leaks would be far from the catastrophes they are generally made out to be. Furthermore, this is actually a problem that will diminish and disappear over time. Radioactivity, by definition, means that the radioactive element will eventually decay and transmute into another element. In the end, only non-radioactive elements will remain. The more radioactive and therefore more dangerous an element is, the faster it will decay and disappear. Elements whose half-lives are measured in millions of years are not very dangerous, as they are not very radioactive.

All this means that radioactive waste is a much more temporary problem than is commonly believed. It is true that even one hundred thousand years from now, spent nuclear fuel will contain some dangerous elements. But it is equally true that after only about one thousand years from now, the waste will not be much more dangerous than a comparable lump of natural uranium. This can be found in abundance from nature: the bedrock above Onkalo repository contains more uranium than Onkalo will ever have, albeit in a much more poorly shielded form. It would be interesting to hear how, exactly, the repository shall endanger people and environment of the far future in ways that the bedrock so far has not?

To compare, there are certain wastes that will remain dangerous practically forever, since they do not have a half-life. These include heavy metals such as cadmium (used in certain solar panels) and mercury that is released from coal-fired power plants. The widely used (and surprisingly widely approved) repositories for these toxins are common landfills and the environment – where we hope they will disperse and dilute. We have yet to see a serious debate on how to handle the electronic waste and toxins that come with solar panels for example, for 100,000 years at least, without any risk at all to the environment and future generations. Or is it okay this time if we just do the cost-benefit and risk analysis with industry-supplied numbers and let the matter rest there?

We already know how to solve the nuclear waste problem

Would you believe us if we said that humans already pretty much know how to solve the "unsolvable" nuclear waste problem, even bypassing the need for very long term storage? There are several already developed technologies that can significantly reduce both the amount and long-term toxicity of current high-level nuclear waste – meaning mostly spent nuclear fuel from current light-water reactors. The most advanced of these technologies is to "burn" the waste in so-called fast breeder reactors. We have more than 400 reactor-years of experience of different types of fast and/or breeding reactors[47]. Russia has been operating its BN-600 since 1980, and its follower, BN-800 in Beloyarsk, started up in 2014. GE-Hitachi has offered the British government its PRISM reactor to help destroy Britain's stockpile of plutonium. According to David MacKay, the government's Chief Scientific Advisor to the Department of Energy and Climate Change, PRISM -type reactors could supply all of Britain's electricity for 500 years just by using their current plutonium and uranium stockpiles.

---

47   Much more information on fast reactors can be found for example here: <tinyurl.com/nreyw8j>.

Most fourth generation reactors have a huge advantage compared to the current 2nd and 3rd generation reactors. They use nuclear fuel much more efficiently – and they can even make (or breed[48]) more fissile nuclear fuel from fertile isotopes that are much more common. Current reactors "burn" mostly the relatively rare fissile isotope, uranium-235, and as a result, can utilize less than one percent of the energy stored in uranium. Fast breeder reactors can "breed" more fissile fuel from uranium-238, which makes up about 99.3 percent of all uranium. Some designs can also breed nuclear fuel from thorium, which is 3 to 4 times more common than uranium. Fast breeder reactors can also use depleted uranium (left over from current enrichment processes), used nuclear fuel (currently stockpiled in temporary storage), weapons-grade uranium and plutonium (which we really want to get rid of as much as possible). There is also a lot of thorium lying around, as it is an almost useless by-product of mining certain rare earth metals. These stockpiles contain enough energy to power the world for centuries at current energy consumption levels. For almost all practical purposes, breeder reactors can be considered to have inexhaustible fuel reserves – which is why some institutions, like the Brundtland Commission originally responsible for defining the term "sustainable development," have for decades viewed them as renewable energy sources[49].

These reactors also "burn" the long-lived isotopes such as plutonium, which are the main worry in storing high-level nuclear waste. They produce a smaller amount of waste that – while very radioactive and containing such dangerous elements as cesium and strontium – will become practically harmless in 300 years[50]. This simplifies immensely the arrangements for long-term storage. In addition, many radioactive isotopes present in breeder reactor waste have valuable medical and industrial uses.

---

48 Reactors that make, or breed, more fissile fuel (for example Pu-239, U-233) from fertile isotopes (U-238, Th-232) than they use, are called breeder reactors.

49 Our Common Future: Report of the World Commission on Environment and Development, chapter 7 (1987), <www.un-documents.net/ocf-07.htm>.

50 Cs-137 and Sr-90 both have a half-life of around 30 years. After 10 half-lives there is only about 0.1 percent of the element left.

The scale of our different energy options is important to recognize. To produce **all the energy needs**, direct and indirect, of a person living a **high-energy life** such as people who live in most developed countries, a PRISM-type reactor would produce an amount of **waste that would fit in a wineglass**. This dangerous waste would disappear in 300 years and would be relatively simple to store for that time. It could not be used to build a nuclear bomb, although theoretically it could be used in a "dirty bomb", where conventional explosives are used to spread harmful material to the environment. There is no other technology that can come even close to such a small environmental footprint. The nuclear fuel reserves now known would be sufficient to power 10 billion people with relatively high living standards for thousands of years. The amount of uranium in seawater, which is constantly being replenished by erosion, would grow these fuel reserves by at least a factor of one thousand. The technologies needed to separate uranium from seawater are already being developed. With breeder reactors, the worries of us running out of uranium, or even facing a "peak uranium" any time within the coming millennia are most likely without substance.

As mentioned briefly above, uranium is not the only choice for nuclear fuel. Another very promising option is to use thorium as a fuel in, for example, a molten-salt reactor (MSR). These types of reactors need further development, but the basics have been known and tested already in the 1960's. There are many recent research projects and start-ups in China, India, the United States and elsewhere. These seek to commercialize thorium as a nuclear fuel and MSRs as a reactor design.

If these designs are so great, then why don't we have these reactors already? It is a good question. First of all, we need to dampen the enthusiasm a bit and note that the best reactors have always been paper reactors. Once projects proceed from plans and prototypes to actual hardware, unexpected and unpleasant surprises crop up almost inevitably. The promoters of fast/breeder reactor technologies and thorium reactors are confident that they can solve any remaining problems while keeping costs down; but we must remember that renewable only-advocates are saying the exact same things.

Breeder reactors also pose some difficult challenges for managing the nuclear fuel cycle: for instance, their startup typically requires large amounts of plutonium or highly enriched uranium[51], and there is a legitimate reason to be concerned how diverting especially the latter to nuclear weapons would be prevented[52]. Theoretically at least, some breeder reactor types could also be particularly useful for countries desiring to produce plutonium for nuclear weapons. Such designs need strong international safeguards. However, certain reactor designs such as the PRISM are designed to be inherently resistant against nuclear materials proliferation, even more so than current light water reactors. There have been no known cases of nuclear weapons material being produced in a civilian power-producing reactor. For someone seriously interested in nuclear weapons, building and operating a simple "nuclear pile" designed to produce only weapons-grade plutonium is much cheaper, simpler, and easier to conceal[53].

In general, the oft-stated belief that civilian nuclear power and nuclear weapons are intimately connected is not supported by evidence. The amount of countries possessing nuclear weapons has been declining steadily, even as civilian nuclear power has been spreading to new countries. For example, Sweden and Switzerland had nuclear weapons programs, until the early 1970's in Sweden's[54],

---

51   The enrichment needs to be around 20 per cent. While this is higher than the 3 to 5 per cent for common light water reactors, anything below 20 per cent is still considered as low-enriched uranium (LEU). Weapons grade uranium needs to be 85 per cent or higher. More information for example here: <mitei. mit.edu/system/files/The_Nuclear_Fuel_Cycle-6.pdf>, pages 93–94.

52   Plutonium required for startup could come from spent nuclear fuel of existing light water reactors. Such "reactor grade" plutonium is very difficult to use for bomb material because of its unfavorable isotopic composition.

53   Among others, such conclusions were reached by the experts involved in the Swedish nuclear weapon project, which seriously considered using civilian reactors for producing weapons material in the 1950s and 1960s. Before the bomb plan was abandoned entirely, the civilian reactor plan was deemed too costly and hard to conceal and defend. In contrast, one of the authors of this book attempted to obtain reactor grade graphite in sufficient quantity for a simple nuclear pile capable of producing weapons-grade plutonium. The first request for quotation was answered through Alibaba.com in 11 hours.

54   Jonter, T. (2010). The Swedish Plans to Acquire Nuclear Weapons, 1945– 1968: An Analysis of the Technical Preparations. Science & Global Security,

and the 1980's in Switzerland's case[55]. One country, South Africa, even dismantled its nuclear weapons while still retaining civilian nuclear power. Even though nuclear weapons proliferation is a serious matter of great concern, it needs to be realized that acquiring nuclear armament is primarily a political decision, not a factor of how many civilian reactors we have. International oversight and cooperation are key factors in stopping nuclear weapons proliferation, as they are in stopping climate change.

Another problem is that new reactor types require a lot of capital to design, prototype and license. A prototype reactor can easily cost anything from a few hundred million dollars to several billion dollars, and few companies have the wherewithal to take such risks. Furthermore, there has been no compelling economic advantage seen for breeder reactors since uranium prices are normally so low as to make fuel costs relatively trivial even for once-through and highly inefficient fuel cycles common to today's light water reactors. Incentives for destroying spent nuclear fuel or other wastes have been minimal, since fuel is cheap and it's technically easy to store spent fuel. In many countries, political and regulatory environments have been risky and lacking clear focus. Any project having "nuclear" in its name faces opposition and large political uncertainties, which translate directly into even larger financial uncertainties.

One hopes that these hurdles are being cleared. Many of the new reactors are designed as modular, factory-built and standardized designs. In theory they would only need licensing once for the design[56], and could be built in an assembly line, potentially reducing costs while increasing quality. All this could help to cut costs in a future ramp-up in the production and deployment of these reactors. However, the political and regulatory difficulties are likely to be formidable. It seems that we are still some way off from the global marketplace where a Chinese firm could license and stand-

---

18(2), 61–86. doi:10.1080/08929882.2010.486722

55   See <nuclearweaponarchive.org/Library/Swissdoc.html>.

56   It would be a huge win for the climate fight if there would be a global standard for licensing reactor designs that most countries would accept. As of now, most countries have their own regulatory and licensing bodies and rules, which makes entering markets both expensive and slow.

ardize a design in China so that it will be acceptable to, say, a British regulator without relicensing in Britain. Nevertheless, the same has already been achieved with airplanes, even though any serious airplane accident tends to kill far more people than the worst nuclear power mishaps.

Furthermore, there simply is not much public discussion about these reactors in the mainstream media. Easily accessible information about these nuclear novelties has been lacking until quite recently, and even many energy experts seem to be only faintly aware of their possibilities. With time, information, public discussion and books on the subject (like the one you are reading, and others[57]), we hope this situation will change. New nuclear technologies are no panacea nor a silver bullet against climate change, but we hope and believe that they could provide valuable services for humanity in providing energy and disposing of the wastes from existing nuclear power plants. Meanwhile, the current nuclear technologies are safe and effective enough to deploy right now.

These views are, of course, not generally shared by the environmental organizations. For the most part, they seem to avoid discussing new nuclear technologies, perhaps because of the fear that they could render moot one of their key anti-nuclear arguments: the waste problem. Most of the time the arguments against breeders, or pretty much any new nuclear technology, circle familiar topics: they have not been and are not economical right now, they are claimed to be dangerous, they are claimed to be untested, and they are seen as a proliferation risk for nuclear weapons material. So if there is any information on breeder reactors, it seems to exist mainly just to confirm how horrible, dangerous[58], expensive and too far off into the future they are, and how they should be banned outright without even seeing what they could do. These arguments can often be heard from the same people and organizations that only five minutes earlier told you to put your trust in clever engi-

---

57 One great example is *Prescription for the Planet* by Tom Blees (2008), which explains and discusses integral fast reactor (PRISM) technology. It can be downloaded for free at <tinyurl.com/9992kma>.

58 In truth, one of the key principles behind fourth generation reactors is that they are much safer, in many cases "walk-away safe", than current reactors.

neers, researchers and capitalists who will no doubt soon invent the affordable technologies that will solve the problems associated with variable renewables, whose high costs are just a reason to spend more money developing and deploying them.

Is nuclear power cheap or expensive?

Key takeaways:

√ In any relevant comparison, nuclear power is competitive with most low-carbon options – which is why the most common argument is that "it's expensive!" and honest comparisons to alternatives are rare.

√ Sometimes one gets the feeling that decarbonizing the energy sector can cost whatever it costs, unless it is done with nuclear, in which case any cost is too much.

√ When comparisons are made, they are usually biased against nuclear in several ways:

√ Comparing name-plate capacity and not the amount of energy actually produced, thus ignoring the higher capacity factors nuclear plants enjoy.

√ Ignoring plant life time, which for modern nuclear is usually 60+ years, with wind and solar having lifetimes of half of that, or even less.

√ Selecting a high discount rate when comparing options. Discount rate is the percentage with which we "devalue" the future compared to the present. A high discount rate gives a disadvantage to investments that are longer term, which automatically hurts plants with longer lifetimes. A high enough discount rate will always prefer consumption now instead of investing for the future.

√ Ignoring quality of energy. Baseload, or dispatchable, power generation is always more valuable to the society than intermittent power. The value of intermittent energy like wind and solar diminishes as their share of energy production grows, as this leads to overproduction at times, with still need for backup-power at other times.

When the award-winning documentary director **Robert Stone** spoke at the Helsinki premier of his latest documentary *Pandora's Promise*, he noted how anti-nuclear arguments have evolved during his career. When he released his first work, the Oscar-nominated, anti-nuclear documentary *Radio Bikini*, the primary force fueling the movement was the fear of nuclear energy somehow destroying the world. Now, the discussion has evolved:

> *"We have gone from 'nuclear is killing our children' to 'oh, we can't afford it.'"*

Discussing nuclear power with many critics of quarterly capitalism, economic growth paradigm and corporate greed has the tendency to turn them into hard-nosed free marketeers *par excellence*. Suddenly, things like corporate profits, investment payback times and short-term cash flows are all-important, and if climate change is to be stopped at all, it has to be profitable to do so – preferably during the next quarter. Since nuclear reactors are not always entirely or immediately profitable in the current energy market, goes the conclusion, they must be banned, lest some foolish executive make the costly mistake of ordering one and thus reducing the global potential for maximum economic growth.

Of course, everything is different about the "true engines" of economic growth, namely renewable energy sources. When discussing these, considerations such as the cost of subsidies they require are entirely irrelevant.

We have already discussed the availability and abundance of renewables, and found out that the more optimistic projections are based on very optimistic assumptions. The 'nuclear is expensive' argument, on the other hand, is based mostly on half-truths. Nuclear power is expensive to build, but only if one does not compare it to other available options. If such comparisons are made, they often are loaded with certain tricks and methods that make the situation look as bad for nuclear and as good for renewables as possible.

The most common methods of distorting the price comparisons include the following.

Installed capacity versus produced energy. It is common to only talk about installed capacity when presenting the prices or the build-up of renewable energy production such as wind and solar[59]. This is not strictly "wrong" in any sense, but it leaves out some very important information. That information is the actual amount of carbon-free energy that a plant produces, and therefore the amount of carbon emissions that are (or could be) avoided. The amount of energy produced is calculated by multiplying the installed maximum (or name-plate) capacity with the capacity factor. The capacity factor is a number between zero and one (or a percentage), with one (or 100 %) meaning that the plant runs at full capacity all the time.

Some real world average capacity factors in percentages are[60]:

- *Coal 50 – 70 %*
- *Nuclear 80 – 90 %*
- *Solar photovoltaics 10 – 20 %*
- *Concentrating solar thermal power (CSP) 30 – 40 %*
- *Wind 20 – 40 %*
- *Hydro 40 – 50 %*

These large differences in capacity factors have major implications to the prices of energy produced when comparing them to the prices of installed capacity.

Ignoring plant lifetime. To continue from the example above, it is not just about the annual amount of energy produced that is important. It is the energy produced during the operational lifetime of plant. Most of the nuclear power plants being built now have a planned lifetime of 60 years, and with proper care and mainte-

---

59   A recent example by Bloomberg with the headline "Fossil Fuels Just Lost the Race Against Renewables" where the author confuses power with energy production. When corrected, the situation looks quite different from the headline, <tinyurl.com/k6urudg>.

60   Numbers mainly from the United States. Individual plants can have much lower or higher capacity factors, depending on many things. The records for well-kept nuclear power plants in the U.S. and Finland for example have been 95 per cent or even higher in recent years, <en.wikipedia.org/wiki/Capacity_factor>.

nance could theoretically last up to 80 or perhaps even 100 years. The economic lifetime of a wind power plant is estimated to be 20 – 30 years, and solar photovoltaic panels are generally believed to perform economically for about as long, although if one is willing to accept reduced performance they might be able to soldier on for longer. So when compared to a similar amount of actual generation capacity (nameplate capacity times capacity factor), a nuclear power plant has roughly three times longer lifetime, and will produce three times as much energy with one investment. Other way to think about this is that wind and solar need to be scrapped and built again three times during the lifetime of a nuclear power plant. The renewable capacity we currently have installed and are building now will mostly not be with us in our 2050 energy production mix.

Selecting the discount rate. It is problematic to put a monetary value for something that is produced far off into the future. To overcome this difficulty, economists and investors use what is called "discounting", which is done by setting a percentage rate by which we discount or devalue the future in comparison to today. To put it briefly, and from the viewpoint of society (not the investor), the discount rate defines how much value we put on the future generations' quality of life and ability to consume. When comparing the value of a plants' total energy production, it is of essential importance how we discount the value of future production. If we select a high discount rate, say ten percent or higher, the energy produced twenty or more years from now has practically no value for our current comparisons. If our discount rate is low (perhaps less than 2 percent), zero, or even negative, we value the energy produced 20 years from now much higher. From society's point of view, the discount rate is fundamentally an implicit ethical choice. Anyone who is worried about future generations and, for example, worsening resource scarcity should implicitly favor a discount rate that is close to zero. With a high discount rate, long-lasting investments that generate benefits far into the future – such as a nuclear power plant – will show higher costs, while short-term investments will seem to be of better value. With a high enough discount rates, it becomes always preferable to maximize consumption today and to postpone investments indefinitely.

Financial institutions such as banks give out loans and financing based on economic calculations that generally assume fairly high discount rates. However, mechanisms such as governmental loan guarantees could be used to indirectly lower discount rates (that is, cost of capital) for projects that are deemed valuable for future generations. The previously mentioned literature review commissioned by the Friends of the Earth UK concluded[61] that if nuclear plants could get financing with the 3.5 percent discount rate recommended by the UK Treasury for decisions that have longer term societal implications (such as decisions about climate change mitigation), the cost of nuclear electricity could in theory be roughly halved. Although the authors rightly note that this is not a realistic way of appraising nuclear economics, it is illustrative of the power of the discount rate.

Ignoring the quality of produced energy. A modern society is dependent on a constant, high-quality energy supply. We prefer to use energy when we need it instead of when it is available. This is one of the fundamental reasons why humans started to invent ways to use external energy sources in the first place – they wanted to break the hold that natural cycles and weather had on their activities and well-being. We need to get warm when it is cold, hence firewood. We need to have light when it is dark, hence modern lighting. To again simplify a bit, energy sources that can meet our needs on demand and on a large scale are called dispatchable, and are of higher comparative value to society. Energy sources that vary according to weather are called intermittent or variable, and by definition are of lower value to society. Usually this difference is ignored, and as long as the share of intermittent energy remains low, it is not huge issue: Someone can always find use for the energy generated However, it may become an issue if we were to increase the amount of variable generation significantly.

Ignoring system-wide and external costs and the cannibalizing effect. The more intermittent power we have as a share of

---

61   Tyndall Centre, University of Manchester. A Review of Research Relevant to New Build Nuclear Power Plants in the UK. Including new estimates of the CO_2 implications of gas generating capacity as an alternative. A research briefing commissioned by Friends of the Earth England, Wales and Northern Ireland, 2013. Page 37.

total production, the costlier it will be to manage. This is because increased amounts of variable, intermittent power sources induce increasing system-wide costs. This escalating cost is usually ignored, even by people who seriously propose 100 percent renewables. Such a target implies that a lot of the electricity generation will indeed be intermittent, or the definition of "renewable" may have been used extremely loosely. It is often assumed that building more solar and wind will be as valuable no matter how much existing solar or wind generation is connected to the power grid. Using such an assumption, one can conclude that if the cost of 20 percent renewable penetration is – say – a billion dollars, the cost of 100 percent renewables would be merely five billion dollars. Unfortunately, this is not the case.

A simple rule of thumb for this is that the smaller a capacity factor a plant has, the smaller share it can have before system costs start to rise precipitously and the value of generated electricity will fall. Because these plants often mostly produce at the same time on a country-wide level (for example, when the sun is shining; note also that weather patterns that largely determine wind conditions can persist over surprisingly large areas), the electricity they can produce becomes less and less valuable as their share of total production increases. If generation exceeds demand, the price generators receive will fall. The price may even be negative, if the grid operator has to pay someone to take the unwanted electricity out of its hands. New production capacity thus cannibalizes the value of the existing capacity of the same type. This also makes reaching "grid parity" a rather meaningless target, as this parity will keep moving away at the same time as installed capacity increases[62]. Of course, if there is a feed-in tariff in place for such production, the producer isn't necessarily concerned whether the produced energy has any value.

As a simple rule of thumb, the maximum economically sensible penetration of any energy source in the electricity grid seems to be roughly equal to its capacity factor. The value of the electricity produced will however start falling long before this level of penetra-

---

62   Grid parity for an electricity source means that its costs are equal to or lower than current market prices.

tion is reached, depending on the grid and availability of supporting load-following capacity like hydropower. Even with foreseeable increases in storage capacity, technology and demand flexibility, it is our opinion that the growth of wind and solar PV generation in a Europe-wide grid (for example) is likely to slow and stall at below 40 and 20 percent, respectively. Feed-in tariffs will become increasingly expensive to maintain, which will eventually erode the political support they now enjoy.

When these factors are included, nuclear power is quite, if not very, competitive.

Ultimately, with all the fuss and hand waving about solar and wind being so cheap and competitive, it is puzzling that they are generally speaking not being built without hefty subsidies or feed-in tariffs. At the same time, the very same hand wavers often complain that nuclear investments are drawing capital away from renewable investments. Finland is an interesting empirical proof to the contrary: even though two new nuclear power plants are being built[63] (without any direct subsidies) in this country of mere 5.5 million people, the wind power sector in Finland overheated in a few years to the extent that even the wind industry insiders became worried. This happened because of rather generous feed-in tariff (83.50 €/MWh for 12 years) that attracted investors from around the world. If nuclear power were to gobble up all the money from renewable investments, this shouldn't have been possible. We believe the more likely explanation is that what the world is lacking is not money, but profitable investment opportunities.

It remains somewhat of a mystery how exactly nuclear investments are taking away from wind or solar investments, for which there seems to be loads of capital available as soon as tariffs are high enough. Will there be more taxes for the government to spend on tariffs if nuclear power plants (and the resulting industry that would use the energy) are not built? Not very likely. It would be

---

63   Olkiluoto 3 at 1,600 MWe is scheduled for 2018 startup, Fennovoima's Hanhikivi 1 at 1,200 MWe applied for construction permit in 2015 with schedule for 2025 startup. Olkiluoto 4 did not apply for construction permit in 2015 due to OL3 reactor still being built, and Fortum's plans for Loviisa 3 are currently uncertain.

possible that some of the money intended for nuclear would go to renewables instead. But how much? And would some of it go to fossil fuels instead, making the whole thing a disservice to the environment? The argument that nuclear investments compete directly with renewable investments is problematic on many levels. First, it assumes that nuclear investments would be directed wholly for renewables if the nuclear plant is not built. Secondly, it assumes this is done without subsidies or tariffs. Third, it assumes these investments would at least be as cost-effective a way to produce low carbon energy as the nuclear plant would have been (this might or might not be true, depending). Fourth, it further assumes that we for some reason cannot build both nuclear and renewables at the same time.

All this comes down to the total cost of deep decarbonization, and how to keep it as low as possible. While this is a really hard and complicated question, there have been some studies on the matter. A recent report[64] from the United States concluded that 80 percent decarbonization done mainly (but not entirely) with renewables would cost around four times as much as decarbonizing with mostly nuclear power. As common sense would tell us, if we are not allowed to use all the tools we have, the result will most probably be both worse and more expensive than if all the tools were allowed.

So do we want expensive or cheap energy?

Discussions about electricity (or energy) prices are often full of internal contradictions. First, people – often influential politicians – use low electricity prices (or predictions thereof) as a reason to oppose nuclear energy investments. This assumes that those who are doing the investing simply have no clue what they are doing, so

64 Williams, J.H., B. Haley, F. Kahrl, J. Moore, A.D. Jones, M.S. Torn, H. McJeon (2014). Pathways to deep decarbonization in the United States. The U.S. report of the Deep Decarbonization Pathways Project of the Sustainable Development Solutions Network and the Institute for Sustainable Development and International Relations. www.deepdecarbonization.org/ and Charles R. Frank, Jr (2014). The net benefits of low and no-carbon electricity technologies, The Brookings Institution, <tinyurl.com/l2z2v8g>.

someone else has to decide how they should use their money[65]. But if electricity prices are low, that means that no other investments in clean energy are going to be profitable either – so should we oppose them as well? Low energy prices can be either good or bad, depending mainly on the reason why they are low. If they are low because the producers have externalized their costs to society – like with particulate pollution and greenhouse gas emissions – low prices are bad thing. In this case, the **price** is low, because someone else is paying the actual **cost**. If the price is low even when most of the costs are being included in the price, it is very good news.

In the shorter term, the market prices can also be low because there is too much capacity, but the elimination game is still ongoing. Overcapacity can be caused by a recession, heavy industry moving somewhere else or because of market disruptions like generous feed-in tariffs that have added some kind of capacity even though it was not actually needed. If prices are low, new energy-consuming investments look more profitable, and more expensive power production capacity looks less so.

High prices can also be seen as either a good or a bad thing. They are bad because a larger portion of the population will suffer from energy poverty and will have more limited options to consume other things than the bare necessities, of which energy is a major part. They are good because they drive investments, R&D and human behavior towards greater energy savings, more efficiency and cheaper ways to produce energy. What is often missed is that basically all of these will tend to lead to lower energy prices in the future. If the energy prices are low, even when most of the costs are included, then there is not much sense in going to great lengths to save energy or invest heavily for small gains in energy efficiency. It would be more sensible to use the same resources somewhere else, than trying to minimize something that is already very small.

So cheap energy is not the problem – if the price includes all or most of the costs. Alas, this is not often the case. This is evident in the disturbing fact that we are burning fossil fuels at an unprecedented rate even though science shows we should leave most them

---

65   This is, of course, what politicians do for a living.

unburned to avoid catastrophic climate change. The take-away message here is that if we intend to do something serious about climate change, the price of (particularly fossil) energy likely will have to go up. This will mean plenty of investment opportunities for clean energy. But if next to nothing is done, prices could remain relatively low. That will mean no investments, at least not to capital-intensive production like nuclear, solar, wind and various energy efficiency solutions.

Electricity, which accounts for just over 20 percent of global energy use, is something we actually know how to make relatively cheaply and cleanly in large amounts. Reasonably cheap electricity should eventually supplement and even replace other energy carriers elsewhere in the energy sector – be it oil used in transportation or domestic and industrial heat made mostly with natural gas, coal and biomass. This substitution absolutely needs to happen. But the requirement for electricity being cheap enough causes huge pressures to build much more low-carbon electricity production – not just to replace coal and gas generated electricity, but to replace fossil fuels elsewhere in the energy sector as well. It is a tall order, but it is also an undeniable fact that has to happen, and it has to happen preferably before mid-century.

There is another possibility, however, and it is very troubling. It is entirely possible that our society essentially decides that new energy generation cannot be built fast and cheap enough to replace fossil fuels without causing societal disruption unacceptable to those in power. In this case, energy prices would remain low, as the true costs of fossil fuels will not be included in their price. The end result would be the continuing use of fossil fuels for much of our energy needs. We have previously shown that climate mitigation scenarios where new energy generation comes only from renewables assume frightfully high build rates and are likely to incur considerably higher costs compared to scenarios where nuclear power is allowed to be a part of the energy mix. The risk of taking this road is the risk described above: that we try but fail to build enough to replace fossil fuels, ensuring that taxes and pollution surcharges on fossil fuels are kept low because there are not enough viable alternatives. In this case, energy prices would remain low due to continuing supply of cheapish fossil fuels. In a vicious cycle, low prices would further discourage investment in more expensive (on paper) clean energy technologies. Eventually, cheap fossil fuels will run out and the prices will rise. But it may well be too late for the planet.

This is exactly what has happened so far, and predictions based on market prices for electricity futures seem to believe this is the most likely future as well. Some anti-nuclear activists have latched on to these predictions in an attempt to argue that nuclear electricity will be unacceptably expensive. But they have failed to notice that such predictions assume that almost nothing will be done to curb the use of fossil fuels. If actions to that effect are taken, nuclear electricity, being on the whole among cheapest large-scale sources of low-carbon electricity, will be competitive at the same time or before electricity from other large-scale low-carbon sources will be.

In general, high prices seem to be problematic only when they can be pinned on nuclear energy. This is evident from reports on the recent British "strike price" system for clean energy, including the Hinkley Point C nuclear reactor. The anti-nuclear movement screamed in frustration when the British government set a strike price for Hinkley Point C nuclear energy at £92.5 per MWh (around 135 € or 140 US dollars at April 2015 rates). This has since become the new official price for building new nuclear power when comparisons need to be made, in spite of the apparently poor understanding of what the strike price system actually means and includes.

What was often left out from the news articles and blog posts were the strike prices promised for other clean energy production. Offshore wind is perhaps the only source with even closely the same scalability as nuclear in Britain, as new onshore wind farms are becoming increasingly difficult to build in this densely inhabited island. The strike price for offshore wind was £155 per megawatt hour. Onshore wind was close to nuclear at £95, while large-scale solar PV had a strike price of £120. Wave and tidal power topped the scales at £305 per MWh[66]. Conspicuously, those using Hin-

---

66 *Investing in renewable technologies – CfD contract terms and strike prices -* <tinyurl.com/pcy9qg3>.

kley Point numbers to argue that nuclear is too expensive without exception compare them to subsidies renewables receive in some other region, never on what renewables in the United Kingdom are receiving. In addition, the contract with EDF, who will be building Hinkley Point C, includes various clawback clauses that limit the profits EDF can make with the arrangement. Similar clauses are absent from renewable contracts. Furthermore, the intermittent electricity produced by renewables is of inherently lower value to the society than dispatchable baseload generated at Hinkley Point.

The reality is that Britain needs huge amounts of low-carbon energy, and they need it fast. The system they are using might well not be optimal, and many consider the British electricity market suffers from a market failure[67], but at least the government is offering strike prices for all low-carbon energy sources instead of favoring just a few.

---

67 As analyzed in his paper *Central Planning with Market Features – How renewable subsidies destroyed the UK electricity market* by Rupert Darnwall (2015), <tinyurl.com/ozagndm>.

High productivity in energy production is a good thing

Key takeaways:

√ The whole history of modern society is a history of increasing efficiency of primary production that has freed more and more people to do something else beyond generating energy. Yet now some people are seriously suggesting that creating jobs in energy production by making it less efficient and more labor-intensive is a good thing. It is not.

√ Jobs that are created with subsidies have two central weaknesses:

√ They disappear when the subsidies stop.

√ They are not added jobs, but shifted jobs, as the money and productivity for the subsidies always comes from somewhere else in the economy – that's the whole reason it's called a subsidy.

√ Facing a future with more automation and less jobs is scary and difficult for many and we need to help with these hardships, but decreasing effectiveness in primary production and putting people to work the fields is not "helping."

If a group of activists from a (fictional) Smallholders' Party should exclaim in public that farming has become too centralized and mechanized, and demand that tractors and harvesters must be replaced with people so more people would get jobs as draft-humans, they would be ridiculed and with a good reason. Mechanizing and centralizing agricultural production have been the main causes of exponential productivity growth in the sector – albeit at the price of increased dependency on fossil fuels, fertilizers and pesticides. This productivity and growth has enabled hundreds of millions to leave the farm and become something else: from factory worker to teacher to dentist to marketing assistant to cultural critic. Some have even become non-fiction authors writing about energy and climate, relying on the work done by thousands of scientists who have also been able to leave the farm. The majority of people no longer have to work the land to feed themselves and possibly a few others, as was almost universally the case just some generations ago. These people freed from the yoke have been able to build the modern society where the key health threat is obesity, not starvation, and standard of living for most far surpasses what the kings of yesteryear could even dream of.

How is it then possible that activists, top politicians and even university professors can propose that we do largely the same thing presented above, but this time with energy production, and it is seen as a great idea? Suddenly the menial low-productivity jobs, largely born from the sheer inefficiency of the suggested model, are seen as nothing but positive. Everybody seems to disregard the fact that if jobs are created with subsidies and if they exist largely because of inefficiency, then they are jobs that are taken away from somewhere else. That is, less people teaching, less working in health care, less creating art and culture, and more people working in primary energy production.

Our current society, where most of us work in some other business than primary energy production (which includes agriculture), is only possible because we produce food and energy efficiently. We have a big surplus of production per person, so one person can feed or provide energy for many others. Those others are then able to do something else. Less efficient primary production will therefore have a job-creating effect by design. But if these new workers in their new jobs need to get paid, it will increase the price of energy, whether we measure the energy in calories or megawatt hours. More expensive necessities, such as energy and food, will leave less available income for other purposes. Reduction of discretionary income will effectively destroy jobs in those non-essential sectors, like in many services. Cheaper energy works in the opposite way, leaving more income to be spent elsewhere to create and sustain jobs.

Another argument is that domestic energy production, even if less effective and more expensive, could replace imported energy.

This would leave the money circulating inside the region, which helps to create jobs for those currently unemployed. The net effect depends on the price differences between imported and domestic energy. To simplify, more expensive domestic energy adds local jobs and keeps the money circulating locally, while less expensive imported energy takes part of the money out of the local economy but leaves more money for other consumption, which also creates jobs. In a market-based system, the less expensive option is usually preferred by the buyer.

Of course, it is a difficult situation for those facing job loss as the world keeps changing. This constant change is a real problem (and a huge opportunity), and it has been one since the early days of the industrial revolution, as most of us have grown with the idea that one needs to have a job and earn a living. Our society is also built on that premise.

Unfortunately, we seem to be far away from a society where robots and machines do all the mundane and dangerous work for everyone's benefit instead of just the robots' owners' monetary gain, and where us humans are mostly left with other pursuits. It might be a good thing to get people to work, if just to keep them occupied and make them feel useful. In some cases, there might be other arguments, such as increasing energy security or cutting energy imports' share in the national trade deficit. But it is one thing to practice social policy with make-do work placements, and another to say (even if implicitly) that reducing productivity in some sector is suddenly a great thing. It is not.

The mixed blessings of do-it-yourself energy

Why is the possibility for people to produce their own energy, while still being utterly dependent on the built infrastructure, seen as a great thing? Of course there are real positive things to be said about this trend of self-produced energy, but they are often not discussed. In our thinking, one of the key positives is the fact that when people start to produce some of their own energy, they also tend to become much more aware of their general energy use and often begin to reduce it.

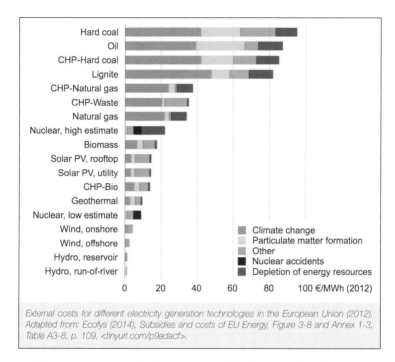

External costs for different electricity generation technologies in the European Union (2012). Adapted from: Ecofys (2014), Subsidies and costs of EU Energy, Figure 3-8 and Annex 1-3, Table A3-8, p. 109, <tinyurl.com/p9adacf>.

More common in the "do-it-yourself"-discussion is to hear arguments that go more like "produce your own energy, and you don't have to pay energy taxes!" Yes, we see how this will help sell more photovoltaic panels, because people rarely like paying taxes. But when this argument is heard from supposedly progressive organizations, who in other matters generally tend to identify more with the tax-loving left than the tax-hating right on the political spectrum, it makes one think. In our view, paying taxes, and getting those taxes back as services from the society, is both morally sound and essential to the functioning of a just and egalitarian society. Those taxes left unpaid are taken away from somewhere else. It is a classic case of the tragedy of the commons, where personal gain is pursued because everybody will share the cost of doing so.

The negatives are almost never mentioned. Some solar panels on one's roof will not make one independent from the grid. Someone still needs to maintain the grid and incur the actual costs, which by and large are not dependent on the amount of energy transferred.

A grid connection used only during darkest days of the year costs about the same to build and maintain and uses about the same amount of materials than the connection used all the time. Since when has it been a wise use of limited resources to build the same service infrastructure many times over if it can reliably be done by building it just once? (If distributed energy generation is a great thing in and of itself, what about distributed mobility i.e. private cars?) Is it really a good thing that the current utilities – upon which we are completely dependent to provide for us when the sun goes down and the wind dies – face diminishing profits and therefore have to cut costs and invest less? Even if one hates the utilities, are we sure that those same utilities won't find a way to make ends meet somehow anyway, with for example hiking up prices they charge for transferring energy or from keeping available some backup power generation in case it is needed? Of course they will, because our society is utterly dependent on constant, reliable energy for the foreseeable future.

Finally, there is opportunity cost. Time, money and resources spent somewhere cannot be spent somewhere else. So even if an investment could be a decent one, it is always wise to do some comparing with other possible investments, consumption, or perhaps saving the money for another day and another opportunity.

Subsidies might be necessary, but their problems need to be discussed

Key takeaways:

√ The major justification for subsidizing some energy sources is the fact that fossil fuels largely externalize their costs and therefore have an unfair and distorting effect on the market.

√ The estimated amount of costs fossil fuels externalize is roughly an order of magnitude larger than subsidies to either nuclear or renewables. Total renewable subsidies soon exceed total nuclear subsidies.

√ Feed-in tariffs alone have proven themselves to be a poor way to subsidize clean energy sources – it is possible that investment grants or R&D support would work better.

So what do we think about renewable energy subsidies? To put it shortly: it's complicated. On the one hand, we will likely need some kind of subsidies to phase out fossil fuels. On the other hand, the current models are often just a money transfer from the poor to the rich that distort energy markets, and they are not likely to promote research and development very efficiently.

The argument in favor of renewable subsidies usually goes like this: They need a level playing field, as fossil fuels are enjoying large direct and indirect subsidies that offer them an unfair competitive advantage. While the exact amount of these fossil fuel subsidies is debatable, this is mostly true[68]. The major, if indirect, subsidy for fossil fuels

---

68 Many oil-producing countries offer cheap liquid fuels to their citizens. This is usually counted as a subsidy to fossil fuels (roughly half of the global fossil fuel subsidy comes from these). What we often forget is that these subsidies are a way for these governments to offer some kind of social security to their citizens by distributing the oil revenues the country gets. It gives the citizens the possibility to consume, just as the social security system in many western countries when someone suddenly loses his job and income. A direct subsidy to gasoline is not the ideal way to handle this, but it is easy to implement in

and biomass burning is the fact that they externalize their costs. Small particulate matter causes sicknesses and premature deaths, carbon dioxide accelerates climate change and so forth. Fossil fuels are cheap to use because someone else usually pays the price.

Of course, many argue that nuclear power also externalizes its costs. While this is true to some extent, the industry in many countries actually has internalized a large part of its costs. The operators collect the nuclear waste, and usually they are obliged to pay a small sum for every kilowatt hour produced to a pool of funds that is used to manage that waste, decommission the power plants and so on. Furthermore, the external costs of nuclear power are not as large as some believe. A recent EU-wide study[69] confirmed that this is indeed the case: depending on the assumptions made about the depletion of uranium reserves, the external costs of nuclear power in the European Union – possible accidents included – are either slightly higher than those of biomass or at about the same level as those of onshore wind. The same study also concluded that the direct and indirect subsidies for nuclear energy were at the same level or significantly lower than those of renewable energy.

Still, the terrible prospect of a nuclear accident and the enormous costs it would entail is often enough to stop people from thinking and making comparisons. Recent headlines have let us know that "The Chances of Another Chernobyl Before 2050? 50%, Say Safety Specialists[70]." The first thing that many readers feel is a terrifying fear of nuclear power, added perhaps with a sudden urge to ban nuclear technology altogether. The study itself is interesting, and we do not criticize here its methods or authors, but we would like to make a comparison.

The study puts the cost of Fukushima accident at $166 billion. This represents full 60 percent of the total cost of all nuclear accidents in a very thorough list, and the authors use six million as the monetary value for a human life. This is an arbitrary choice, but it

---

place of a more complex social security system.

69   Ecofys (2014). Full dataset on energy costs and subsidies for EU28 across power generation technologies.

70   MIT technology review (April 17, 2015), <tinyurl.com/nabdzkf>.

is a value that is used elsewhere as well, and for comparison's sake it is as good as the next one. World Health Organization (WHO) has estimated that burning of biomass and fossil fuels causes around seven million deaths each year. Around half of them are caused by indoor pollution from biomass burning. The total value of this loss of life then comes at **$42,000 billion each year**, with the likelihood of this happening at around 100 percent each year. Even if we account for the fact that fossil fuels now provide roughly 20 times as much of our energy as nuclear, the picture will not change much. These sums do not take into account any sicknesses caused by air pollution and the lost productivity and health care costs it causes, nor the costs of worsening climate change due to increased greenhouse gas levels. Chernobyl is estimated to cause up to 4,000 deaths by 2050[71]. Assuming such accident is even physically possible with currently operated reactors – a major and contentious assumption – an estimated 50 percent risk of another accident with similar consequences by 2050 is of course news, and risk assessments such as these are important. But we need to do the comparisons. Burning hydrocarbons as we do now will have a 100 percent likelihood of killing hundreds of millions by 2050. Lack of energy and the services it provides will kill even more.

Including the external costs in fossil fuel prices would be the preferable, market-based solution to the problem. Emissions trading in its various forms is one way of doing this for the greenhouse gas emissions. But it is still a long and difficult journey, no less so because we do not have many credible and scalable alternatives for fossil fuel use in many places. In the give-and-take of politics, our ability to implement these pricing mechanisms, be they carbon taxes or emission trading systems, is actually exactly as strong as the sum of low-carbon alternatives we have on the table. The less alternatives we have, the less success we will have in limiting fossil fuel use.

To increase the amount of credible alternatives, we probably need subsidies. We need many different sources of low-carbon energy to scale the total capacity faster. We also need to do experiments – such

---

71 See for example United Nations press release here: <www.un.org/press/en/2005/dev2539.doc.htm>.

as what Germany is currently undertaking on a large scale – with large amounts of intermittent energy sources coming into the grid to find out if they are manageable and at what cost. We cannot know these things from models and calculations alone. But we need to have an honest and open discussion about subsidies and their effects. The subsidies and energy prices have to be compared fairly and in relation to the amount of energy produced. For example, in Germany the absolute amount of various subsidies that nuclear power has received in that country is currently still slightly larger than the amount received by renewables, but when compared by the amount of energy produced, renewable subsidies are roughly four times larger.

Another discussion needs to be had on the regressive nature of many subsidies. In practice, much of the wind and solar feed-in tariffs in Germany have resulted in an enormous wealth redistribution from the poor to the rich. The rich have the available land, roofs and capital to invest in photovoltaic panels for example. Around two percent of German households have solar panels, and most of Germany's solar power is produced in installations that have cost at least 200,000 euros. Smaller installations, costing up to ten thousand euros[72], produced around ten percent of all solar electricity, and therefore received around ten percent of the tariffs. For the wealthy, solar panels have been a risk-free investment with government-guaranteed returns well above the riskier stock markets: a no-brainer for anyone with the option to participate.

In theory, even those with smaller fortunes can participate through different co-operatives, but studies show that these co-ops fund no more than ten percent of the annual investments. The vast majority of the members in these co-ops are middle-aged, wealthy white males[73]. In practice, it seems that the tariffs are a money transfer to big industry, which is relieved from the EEG surcharges, to rich investors, and to the wealthier upper middle class. Is it any wonder, then, that Energiewende is so popular amongst the middle

---

72   According to this paper by Bernard Chabot: <tinyurl.com/px84bqv>.

73   Özgür, Y. et al., (2015), Renewable energy cooperatives as gatekeepers or facilitators? Recent developments in Germany and a multidisciplinary research agenda, doi:10.1016/j.erss.2014.12.001

class? The system in place is practically a wealth transfer from one part of society to another, and it is rather misleading to call this a "democratization" of energy production, as is often done.

One of the stated goals of the Energiewende and similar renewable-boosting plans is to cut carbon emissions. Are feed-in tariffs an effective way to do it? The EU, and several other areas in the world, have emissions trading systems (ETS) in place to limit emissions. In short, emission trading means that a fixed quota of emission permits is sold, auctioned or given away to polluters. If you want to pollute more, you have to buy more permits. If you are able to decrease your emissions, you can sell your permits to someone else. ETS is a market-based solution that should – left to its own devices and unless drilled full of holes by lobbyists – cut emissions in the most cost-effective way. However, when combined with national renewable feed-in tariffs and the economic depression, the EU ETS is currently not working as it should. The price of carbon is far too low to incentivize investments in low-carbon energy. The quantity of emission permits available is simply much too large. Apparently nobody foresaw the recession or the fact that national schemes that reward the builders of new electricity generation irrespective of whether someone needs the electricity would distort the market and make cost-effective investments less appealing.

In addition, simple feed-in tariff systems can slow down research and development of technology. When profits are guaranteed, it is safer to just build the current technology and collect the tariffs reliably than to venture into new technologies with all the risks involved in innovation. Innovation researchers argue that subsidies should be directed more towards research and development even at cost of reducing feed-in tariffs, and we tend to agree – although with certain reservations. In the context of climate change, we are ultimately going to need energy sources that can outcompete coal. Without such energy sources, pressuring poor countries to forgo fossil fuels in their drive to develop could prove to be impossible. The goal of having energy sources that can routinely beat coal in its own game may seem far away. If we leave out nuclear power, the one energy source that has proven history of being capable of outcompeting coal plants, the goal will be even farther away.

At the end of the day, the misleading rhetoric and overtly optimistic hype around renewables could lead to a disappointment and severe loss of trust in them. Already there are increasing problems with local opposition to wind farms. The tariffs, even with all the "democratization" spin around them, are already causing bitterness and feelings of inequality. Critics who raise these issues to public discourse are often met with arrogance. This is exactly how nuclear boosters treated their critics in the 1970s, and the end result is all too likely to be exactly similar as well: eroding public support and disappointment in entire field of technology as renewables turn out to be just energy generating technologies among others, not magical silver bullets that solve all our problems at once.

What can we still do?

We need all available options

The key message of this book bears repeating: if we want to halt dangerous climate change, we will have to use all the means available. We need strong policies and actions to **promote renewable energy, energy conservation and efficiency, nuclear power and carbon capture and storage**. We also need to find ways to replace fossil fuel use in industry, and to start reforesting the planet in earnest.

It seems likely that these actions, as recommended by the IPCC, will not happen fast enough. The longer it takes for us to get our annual emissions to a permanently declining trend, the harsher, more expensive and therefore less likely changes are needed. We are quite likely already on a trajectory where we also need to start thinking about geoengineering solutions. Geoengineering poses huge unknowns and risks, and can demand large amounts of energy and resources. They should be an emergency solution, to be used only if we fail even with all the low-carbon energy options at our disposal.

In the first part of the book, we criticized the optimism some pro-renewable, anti-nuclear activists seem to have. Evidence shows that stopping climate change is not as easy as many of these people and organizations seem to believe. Either they have not understood the scale of the problem, or they do not want to talk about it. We understand the need for optimism and positive thinking, but at the same time, uncritical optimism can be misleading, even harmful. When the public is led to believe that we have plenty of solutions for the climate problem, even the small pressure to actually do something about the problem right now could evaporate.

There is lot to we can do. As a first step, we urge everyone who consider themselves environmentalists to rethink their prejudices and priorities. Read the latest IPCC reports (not just the catchy news headlines) and decide what is the most important battle and our most important goal. The prevailing rhetoric that often frames renewables as a way to replace nuclear power is a major asset to the fossil fuels industry. Pro-nuclear and pro-renewable people need to join forces to tackle the climate problem, which is mainly caused by the burning of fossil fuels. We absolutely need all workable solutions right now.

Nothing in this book should be construed to suggest we should become uncritical supporters of nuclear power and industries behind it. There is a place for nuclear criticism, as there is one for criticizing various renewable energy sources. We are not asking people to stop thinking or voicing their opinions. What we ask is for a prudent use of limited resources. The main goal should be to halt climate change as fast as possible. Direct your activism and energy accordingly; block new coal plants and seek to close those still operating, support Greenpeace and others in their efforts to block Arctic oil drilling, and tell your elected representative that we need – and want – a higher price for carbon emissions, locally and globally. But remember the reality: we are likely to drill oil and burn coal and gas as long as there are not enough viable options available elsewhere. They are simply too valuable to pass up on unless we have enough viable alternatives. This is the reason why opposing planned nuclear power plants often plays straight into the pockets of the fossil fuel industries. As the saying goes, perfect is the worst enemy of good enough. We need to see the big picture, where opposing one alternative because of its imperfections often leads to another, even worse solution being adopted instead. All the "good" options suffer from major scalability issues; there are limits to how fast we can manufacture solar panels, wind turbines and nuclear reactors, and there are limits to how rapidly we can increase the manufacturing capacity. In this context, replacing one low-carbon energy source with another is simply a terrible waste of resources.

We also need to keep our eyes on the target. There are some paths we can take that seem to reduce carbon dioxide emissions but will never take us to our goal of carbon neutral society. We can replace coal with natural gas and biomass in electricity production and cut emissions – at least on paper – quickly and efficiently. We see this happening in the United States, where the fracking boom has dropped natural gas prices so low that utility companies are building gas-fired power plants and shutting down coal plants. But these solutions will never take us all the way. Because of the above-mentioned scalability problems of zero-carbon energy, they can be used as stepping stones to buy some time where zero-carbon energy is not currently a viable solution. But again, it would be very counterproductive to oppose nuclear power or renewables just because we can have **some** emissions reductions cheaper and even faster by taking a step down a road that will never take us the whole way.

If we need to get to the Moon, it would be a poor use of resources to start by building a skyscraper to get a little bit closer. We need to start building the rocket straight away with all the resourcefulness we can muster.

We also need to be honest about the difference in reducing emissions on paper and actually reducing emissions. The IEA has warned that natural gas, due to leakage of methane, could be as polluting as coal. It's possible it could be even worse[74]. The fact that biomass is currently seen as carbon neutral is an artefact of the emissions trading and carbon accounting systems used, not a physical fact. Environmental organizations need to be honest in their rankings of countries according to their climate policies. Claiming that nuclear power is as bad as coal or other fossil fuels when comparing climate solutions, just because one doesn't like nuclear, is wrong. Lies will not only lead us in the wrong direction – these reports often suggest that it is good policy to replace nuclear with anything but coal – but will also eventually destroy the trust in these organizations.

We have to stop using energy conservation, efficiency, wind power or other good solutions as blunt rhetorical instruments to oppose nuclear power. We most probably are going to need all these solutions, and we very probably need nuclear power as well. Using all of the above together, we might just have a chance at stopping climate change fast enough. Similarly, those nuclear advocates that oppose renewables must realize that renewables are here to stay, and will be providing a good portion of world energy use in the future. It is no use to spend time and energy arguing with people who at least share the vision of low-carbon society.

The authors of this book are not the only ones hoping for these priority adjustments. Four of the leading climate researchers and experts wrote an open letter in 2013 to the environmental organi-

---

74   For more on this, see for example this paper: <tinyurl.com/nwjsrck>.

zations of the world, hoping they would reconsider their opposition to nuclear power[75].

If that actually happened, and if even one major environmental organization would announce that the climate crisis is causing it to call a ceasefire against nuclear power, it would send a huge message about the urgency of climate change mitigation. So far, the anti-nuclear dogma has made climate change communication much more difficult and ineffective. One wonders at the dubious logic that permits these organizations to oppose our most important source of zero-carbon energy at the same time as they tell us we have to stop emissions now. We know for certain that this has caused many people to believe that climate change itself is not a problem. It seems likely that most of us have to give up something in order to have a chance at halting climate change. Would it be too much to ask the anti-nuclear environmentalists to at least consider giving up their opposition to nuclear power?

Are environmental organizations part of the problem or part of the solution?

As it stands, it seems that environmental organizations, due to their anti-nuclear heritage going back to the 1970's and 1980's, could have a net-negative effect in climate change mitigation. Even if these organizations do a lot of valuable work in many areas, a sizable amount of their resources goes to opposing nuclear power. In reality, this means opposing a major source of low-carbon energy. Since these organizations lack the means to force us to build other low-carbon energy sources instead, the likely result is the building and maintaining of the current fossil fuel-based energy sources. This is a significant problem. If even a relatively small part of resources that would have gone to building nuclear are actually used to increase the use of fossil fuels – or even biomass – the climate fight gets harder. In reality, it is very likely that a significant portion of the resources that would have been used to build nuclear power will indeed trickle to dirtier energy sources. This means that despite

---

75   Letter can be read for example here: <tinyurl.com/o63s6k7>.

their best intentions, the net result of the policies put forward by these organizations can be bad for climate.

Can these very large and already established environmental groups change? Alas, maybe not. From experience, we know that large organizations often become myopic to the changes in their environment, and fall victim to group-think that stifles internal criticism. These problems are even more likely to arise in strongly ideological organizations, such as political parties and environmental organizations, particularly so when many of their activists are volunteers bound together by ideology instead of financial obligations. A new member in these groups is expected to share the ideologies and values of the organization and other members. In an echo chamber where very few people have – or dare to voice – dissenting opinions, existing opinions tend to harden and radicalize. Those who disagree are either discouraged from joining the organization in the first place, or are likely to be smoked out after a while. As a result, the organization's views tend to harden even more. At worst, those thinking "wrong thoughts" can even be seen as turncoats and enemies.

Let's not forget that many environmental organizations are also very big businesses, some with annual budgets of hundreds of millions of dollars. Stoking fears about nuclear power has been a staple of their fundraising efforts for decades. So like any large organization, their leaders must be considering what might happen to their fundraising efforts if they suddenly began supporting nuclear power. Might they lose a substantial portion of their members (and financial support)? It seems that the years of fear-mongering have painted them into an anti-nuclear corner, whether they actually want to be there or not.

While researching and writing this book, we have had numerous private discussions with members of the Green parties and environmental organizations. We have learned that surprisingly many of these people actually think the official policy of their organization towards nuclear energy is misguided and lacks solid evidence. Many "greens" already support nuclear; in Finland, the percentage hovers at around 30 percent of the Green voters. Even if they do not endorse everything about it (as no one should), then at least they support it as a "lesser evil" in the climate struggle, or as a possible future energy source should better reactor designs emerge. This is very encouraging. On the other hand, we also know what an uphill battle these people face. They put their social status and even their relationships in jeopardy if they voice their opinions or demand change too vocally. In some cases, even their jobs may be in danger. We are personally aware of cases where a positive attitude towards nuclear energy has resulted in an end of a friendship. It might not be far-fetched to think that a career in an environmental organization would face unexpected obstacles, if an employee "comes out" with his or her positive attitude and arguments in favor of nuclear power, let alone states something publicly on the matter.

If the current environmental movements cannot change, we will eventually need a new kind of environmental movement. There are signs of this already happening. More and more environmentalists regard nuclear power and technology in general as tools for positive change, if we have the will and wisdom to use them as such. They also often view rising living standards amongst the poor as something to be embraced and welcomed, as they see the positive effects it will have for human dignity, gender equality, fertility rates and even environmental protection. Seeds of new environmentalism have already been planted: a movement calling themselves Ecomodernists recently published their *Ecomodernist Manifesto*[76]. We warmly recommend reading the document with an open mind and considering whether one could support the thinking behind it. Many have already: an ecomodernist association has been founded in Finland in June 2015, and there are activists in many other countries. Of course, there are also critics of the "new environmentalism", especially amongst the old guard of environmental thinkers. The main problem we have with this critique is that despite decades of theorizing and trying, advocates of more traditional environmental ideals have not succeeded in getting people at large to accept the needed cuts in living standards and well-being upon which their visions depend, nor have they come up with a way to force this development. In contrast, the Ecomodernist Manifesto paints a more

---

76   See <www.ecomodernism.org/manifesto/>.

optimistic picture of the future, one that might be much easier for the general public to accept.

What can you do?

Climate change is a big issue. It is so big that it cannot be left to individual citizens or "consumers" to deal with. We need a functioning, efficient climate and energy policy, and to get that, we need citizen activism. Do not surrender to cynicism; raise your voice for the environment and our future. Demand tighter emissions control and more zero-carbon energy – be it nuclear or renewable. Keep your aim on the target, which is to cut local and global emissions and end fossil fuel use as quickly, efficiently and humanely as possible. Demanding more zero-carbon, fossil free energy is one of the means of getting there, but it is not the target itself. This bears repeating:

The goal is to halt emissions to a level that climate science demands, as fast as possible.

The primary target is not 100 percent renewables. Nor is it 100 percent nuclear. Limiting our options based on personal preferences will make it harder and more unlikely for us to reach our goals.

Each of us – and we hope you share the message of this book with your friends and family – needs to demand credible energy policy from both environmental organizations as well as political parties. We have the right to know how they plan to have a carbon neutral society and by when. We also need to know about the risks involved, and the likelihood of them materializing, as well as the associated, comparable costs. "We just need more political will" is not an answer; it is a way to evade answering the question. Of course we need political will, but especially in a democracy – the apparatus which is based on the premise of limiting the political will of the ruling class – political will is as limited a resource as is the human capability to stay alive without nourishment.

We also have the right to know what kind of world we are building, and why. What is the expected global energy consumption and how it is distributed in the alternative scenarios? How do they stack up against historical evidence? If the plan depends on new or emerging technologies, we should ask if these are available, what their limits (scalability) and prices will be, and whether have they been tested on a proper scale. With the future of the planet – our only one – at stake, we need to be able to ask the hard questions, even from those who we believe are sincere and mean well. If we do not do that, it will probably be a disservice both to them and to ourselves, as no amount of public relations can trump the laws of physics.

The answers have to be credible, and they have to be based on peer-reviewed, mainstream science. Rejections of solutions, like nuclear power, that are being recommended by expert organizations such as IPCC need to be based on extremely solid reasoning. Despite our dozens of conversations and repeated polite questions, we have not yet been presented with solid reasons to reject nuclear power.

The final question is perhaps the most important. How does a suggested scenario that reaches the goal without using some solutions compare with scenarios that accept all the alternatives? Is it better, faster, or cheaper, and why? Common sense tells us that limiting our options arbitrarily beforehand will, on average, result in a worse outcome. Cost-benefit analysis and fair comparison are essential parts of any study. To put it very roughly: Arguing that a certain energy source is the best one is very easy if one doesn't actually include alternatives in the comparison. The problem with this approach is that we cannot solve the real problems with lies, rhetoric or clever propaganda. As noted physicist **Richard Feynman** once said, "For a successful technology, reality must take precedence over public relations, for Nature cannot be fooled."

The fact is that we all need to pay attention to is the amount of emissions, and the amount of emissions reductions actually happening. Another fact we need to remember is that we need to eventually get rid of emissions in practically all of our energy use, not just from electricity generation. It is good to remember that electricity is only one fifth of our total energy use, even though it emits a relatively larger part of emissions. We will need space heating and warm water, industrial process heat and especially we will need low-carbon liquid fuels or a way to power our logistics and machines without them. All of this will likely demand much more zero-carbon electricity production in the future.

We are not saying that the nuclear industry should not be criticized. It is a business amongst others, and therefore it has incentives to cut costs and to promise more and better than what it can deliver. The projects underway and those being planned need to stay on schedule and on budget better than they have. This goes for the renewable industry as well, but the nuclear industry is perhaps exceptionally dependent on public acceptance. This acceptance and trust must not be endangered by short-term profit seeking.

The nuclear industry faces a difficult public relations problem, which we think it has tried to solve largely by doing nothing. This is somewhat understandable, since anything the industry says or does is often distorted beyond recognition – with the help of the anti-nuclear movement – as soon as it is published. Any errors made by nuclear industry are blown to huge proportions; if no errors can be found, it is seen as a proof that there is simply something they're not telling. The situation seems ridiculous, but right now it is what it is: a person who knows more about nuclear than 99.99 percent of the population is not trusted to say anything correct – unless it is negative or critical – about nuclear, if he or she is or has been (or even could someday be) employed by the industry. Even academics who research nuclear technology or radiation are viewed with skepticism and almost automatically suspected of bias, corruption or even conspiracies. But whenever anyone says something critical about nuclear power, it is widely seen as profound truth. No matter if the qualifications in this case are limited to being a famous actor, an artist, a science-fiction author or a Greenpeace activist with a degree in literature. This is a worrying development, unfortunately not unique to energy discussions. Witness, for example, the rise of the anti-vaccine movement – or climate denialism, for that matter.

The regulatory framework needs to be streamlined in many countries. To take Finland as an example, each reactor demands a politically risky, lengthy and expensive permitting process. This in part means it is currently very impractical to even think about building anything but very big reactors. Furthermore, reprocessing spent fuel and exporting it for reprocessing elsewhere are also banned. With a single piece of legislation, we have created for ourselves a nuclear waste problem and outlawed many useful ways to deal with it. A breeder reactor such as the PRISM would currently be illegal, as it reprocesses its spent fuel to make new fuel. This limits our choices in what to do with the waste in the future. At the same time the law requires that new projects, such as Fennovoima's Hanhikivi 1, have plans ready on what they will do with the waste they are going to produce. There needs to be a serious public discussion on fourth generation reactors and what they have to offer. A low-carbon, dense, affordable and reliable energy source that has enough fuel already available to last us thousands of years is simply a prize that cannot be rejected out of hand. Nuclear needs to be viewed as the "cleantech" it is. After all, what is cleaner technology than a device that uses hazardous waste as fuel, makes it much less hazardous, and produces vast amounts of low-carbon energy in the process?

Similarly, political decisions that limit the options for decarbonization need to be prevented. The magnitude of the climate crisis demands that we cannot arbitrarily limit the world's access to low-carbon energy generation. Currently, the so-called Bonn agreement, a holdover from 2001, places significant limitations on nations that might otherwise use nuclear energy as one of the tools for combating climate change. The agreement excludes nuclear power projects in developing countries from financial help other energy projects could receive, and should be amended at least. Other regulations and legislation in place need to be reviewed with a cost-benefit analysis in mind, and they need to be compared with other industries and energy production. The politicians and officials need to make solving our climate crisis a priority. Their responsibility is to make it as fast, easy and cost-effective as possible, not as slow, bureaucratic and expensive as possible.

Final thoughts

This book was not written because we are pro-nuclear. It was written because we believe in honest discussion about evidence and want to bring some currently missing points of view to the climate, energy and nuclear debate. It is impossible to evaluate the situation clearly if one does not have enough information. Nuclear power is not a perfect solution nor a silver bullet that will solve our climate crisis. Neither are renewables or energy conservation. Pitting these against each other in a situation where over 80 percent of our energy comes from fossil fuels is a folly of monumental proportions. It takes us in the wrong direction. As long as we pit low-carbon energy sources against each other, the true winner will be the fossil fuel industry. Climate change is a threat so serious that we really need to have a calm, honest and evidence-based discussion about the options we have. Ideologies and prejudices need to give room for science and evidence.

We do not expect that everyone who has read this book will take to the streets demanding more nuclear power. We wish that this book would arouse fresh curiosity and a desire to learn more. Perhaps it can open some new discussions, either for or against its message. We certainly do not hold a monopoly on truth, and it is entirely possible that we have made mistakes or will be proven wrong in some of our statements or predictions. All claims need to be reviewed with healthy skepticism, particularly when they seem to fit personal preconceptions.

Perhaps the most important motivation for this book is the short time frame for decarbonization. If right now we were closing the last power plant fueled by fossil fuels, if the energy system would have very low carbon dioxide emissions, and be otherwise relatively harmless, it is entirely possible that we would be demanding the phase-out of nuclear power if the evidence would suggest it to be feasible. But the reality is that right now we do not have any guarantee or evidence that 100 percent renewable energy system is even possible, let alone feasible or sensible. We do not even know if climate change – arguably one of humanity's most serious and difficult collective problems – can be mitigated enough even with all available prospective solutions, including nuclear power. But we do have evidence that so far, nuclear has been the most effective solution for decarbonizing energy production. We also have evidence that the problems often associated with nuclear power are not as great as we are often led to believe.

The evidence so far would suggest that nuclear power is one of the most essential tools we have in mitigating climate change. We simply do not have the luxury of picking personal favorites and opposing everything else, if there is the slightest risk that it will slow down our mitigation efforts. We are already much too late, and we cannot afford to be any more late than we are. We need everything we have, including nuclear, both in its current form and in the form of next generation reactors.

About the authors

Rauli Partanen is an independent writer, lecturer and consultant on the environment, energy, society, the economy and their interrelations. B.B.A.
raulipartanen@gmail.com
Twitter: @kaikenhuippu
Blog: http://kaikenhuippu.com (mostly in Finnish)

Janne M. Korhonen has a master's degree in engineering. He is a PhD student and independent writer and researcher on innovation, energy and the environment. MSc.
jmkorhonen@gmail.com
Twitter: @jmkorhonen
Blog: http://jmkorhonen.net
Follow the discussion on our website at climategamble.net and twitter @Climate_Gamble.

Climate Champions

Climate Gamble was originally published in Finnish in March 2015, and adapted and translated to English in August 2015. This is the second, revised edition with several minor improvements and corrections. The revision was in part made possible by a public crowdfunding campaign that sought to finance free distribution of copies of Climate Gamble during the COP21 climate negotiations in Paris in December 2015. All the funds collected during the campaign have been used either to print books for distribution, or for distribution itself and its related costs.

To extend our heartfelt thanks to those who participated in the campaign, and in interests of full disclosure, the following individuals and organizations have supported our COP21 campaign by 150 euros or more. You can find this list and updates on our full disclosure also from our website climategamble.net. The misunderstandings, mistakes, and conclusions remain our own, as always.

All in all, we managed to collect roughly 9.000 euros for the project. After expenses, we printed out a run on around 5.000 books.

Climate Champions, in alphabetical order:

Analysgruppen
Areva
Arthur de Montalembert
Cre8 Oy
Energy for Humanity
Fennovoima
Finnish Energy
Mathieu Rouvinez
Mikko Muukki
Timo Laaksonen
TVO

There are also some people that we feel the need to thank especially. They have all helped significantly in various ways in our mission of spreading more evidence based energy policies throughout the world. These people are:

Myrto Tripathi
Lauri Muranen
Tuomo Huttunen
Kirsty Gogan
Tom Blees
David MacKay
David Schumacher